AAPS Advances in the Pharmaceutical Sciences Series

Volume 57

Series Editor
Yvonne Perrie, Strathclyde Institute of Pharmacy and Biomedical Sciences
University of Strathclyde, Glasgow, UK

The AAPS Advances in the Pharmaceutical Sciences Series, published in partnership with the American Association of Pharmaceutical Scientists, is designed to deliver volumes authored by opinion leaders and authorities from around the globe, addressing innovations in drug research and development, and best practice for scientists and industry professionals in the pharma and biotech industries. Indexed in

Reaxys

SCOPUS Chemical Abstracts Service (CAS) SCImago

EMBASE

Seema Kumar
Editor

An Introduction to Bioanalysis of Biopharmaceuticals

Editor
Seema Kumar
Drug Disposition and Design/DDT, Biopharma R&D
EMD Sereno Research and Development Institute
Billerica, MA, USA

Editorial Contact
Carolyn Spence

ISSN 2210-7371 ISSN 2210-738X (electronic)
AAPS Advances in the Pharmaceutical Sciences Series
ISBN 978-3-030-99998-8 ISBN 978-3-030-97193-9 (eBook)
https://doi.org/10.1007/978-3-030-97193-9

This Springer imprint is published by the registered company Springer Nature Switzerland AG
The registered company address is: Gewerbestrasse 11, 6330 Cham, Switzerland

Preface

Large molecule (LM) biopharmaceuticals are sophisticated therapeutics that are much larger in molecular size than their chemically synthesized small molecule pharmaceutical counterparts. They are often derived from biological sources through complex biotechnological processes, such as recombinant DNA technology.

Although biopharmaceuticals have been around for several decades, it is only in the last few years that they have started to catch up with small molecule drugs on their share of market success. The advancement in protein and biomolecular engineering technologies has further enabled the emergence of new and innovative biopharmaceutical modalities. Today a broad range of biopharmaceuticals including recombinant proteins, monoclonal antibodies, bispecific antibodies, multi-specific biotherapeutics, antibody-drug conjugates, fusion proteins (e.g., Fc fusion, PEGylated proteins), gene therapies, cell therapies, peptides, and oligonucleotides are entering in preclinical and clinical development.

Unlike their small molecule counterparts, the large molecular size, complex structure, and heterogeneous nature of biopharmaceuticals pose challenges in their full characterization. Additionally, biopharmaceuticals could be seen as foreign biological entities by the human immune system and could invoke an unwanted immune response. Evaluating the safety, efficacy, pharmacokinetics, and pharmacodynamics of these biopharmaceuticals thus requires monitoring of the biotherapeutic molecule and the immunogenicity response against it. Bioanalysis, i.e., quantitative measurement of biopharmaceutical in biological fluids, primarily blood, plasma, serum, urine, or tissue extracts, enables these evaluations and thus serves as an integral part of the biopharmaceutical drug development process.

This book provides a comprehensive overview of the fundamental and practical aspects of bioanalysis and the vital role it plays in the development of safe and efficacious biopharmaceuticals with speed and cost-effectiveness. The book begins with an introductory overview of bioanalytical methods and analytical platforms, progression of bioanalytical strategies through various stages of drug development (discovery, preclinical and clinical), current scientific practices, and evolving regulatory landscape for bioanalysis.

The subsequent sections of the book dwell further into biopharmaceutical modality-specific bioanalysis including bioanalytical strategies, associated bioanalytical challenges and mitigation approaches, industry best practices, and latest understanding of regulatory guidance as applicable to the fast-growing biopharmaceutical landscape.

Billerica, MA, USA Seema Kumar

Contents

About the Editor

Seema Kumar is currently Director and senior DMPK lead at EMD Serono Inc. (a business of Merck KGaA, Germany). In her role, Dr. Kumar leads the Clinical Bioanalytical Sciences group that provides Bioanalysis (BA), Immunogenicity, and DMPK support for clinical stage large molecule portfolio. Additionally as a senior DMPK lead, Dr. Kumar provides overall drug disposition support for large molecules spanning across discovery and development stages of the program. Previously, Dr. Kumar led regulated Bioanalytical group in BioMedicine Design (formerly PDM) at Pfizer. The group provided regulated (GLP/GCP) BA support including assay development, validation, and sample analysis for PK and Immunogenicity (ADA and Nab) assessment of Pfizer's large molecule portfolio. Prior to joining Pfizer, Dr. Kumar served at roles of increasing responsibility as Director of Quality Control and Director of CLIA-certified Clinical Bioanalytical Lab at XBiotech USA Inc.

Dr. Kumar holds a Ph.D. from Johns Hopkins University and has published several publications in peer-reviewed journals and contributed to several book chapters. Dr. Kumar has given numerous invited talks in various national and international scientific conferences and meetings and is an active member of industry consortia (IQ, AAPS, etc.).

Chapter 1
Introduction

Sanjeev Bhardwaj, Inderpal Singh, and Matthew Halquist

Abstract The bioanalysis of biopharmaceuticals is central to the drug development process. The efficacious dose projections of biopharmaceuticals rely on a robust, sensitive, specific, and selective bioanalytical assays that accurately quantitate the analyte(s) of interest. The bioanalysis of large-molecule biotherapeutics is more complex than their lower molecular weight counterparts. The immunogenicity of large-molecule biotherapeutics brings another level of complexity to their bioanalysis. The advent of new biotherapeutic modalities such as bispecifics, multi-specifics, fusion proteins, antibody-drug conjugates, oligonucleotides, gene and cell therapy drugs that span beyond the realms of large-molecule and small-molecule therapeutic modalities has made the bioanalytical landscape more complex. Traditionally, ligand binding assays (LBA) and liquid chromatography coupled with mass spectrometry (LC-MS) have been the work horse for large-molecule and small-molecule therapeutics, respectively. However, the expansion of therapeutic modalities beyond the conventional realms of large-molecule and small-molecule therapeutics has led to the increased utility of combination of LBA and LC-MS approaches for the bioanalysis of these complex therapeutics. The challenges of gene therapy and cell therapy bioanalysis are attributed to their unique delivery systems, technologies, and the complex nature of immune responses against these multi-component therapeutics.

Keywords Biotherapeutics · Bioanalysis · Immunogenicity · ADME/Drug disposition · Pharmacokinetics · Pharmacodynamics · Ligand binding assay · LCMS assay, Intact mass · Trypsin digestion · Signature peptides

S. Bhardwaj (✉)
Janssen Biotherapeutics, Spring House, PA, USA
e-mail: sbhardw4@its.jnj.com

I. Singh
Spark Therapeutics, Philadelphia, PA, USA

M. Halquist
Virginia Commonwealth University School of Pharmacy, Richmond, VA, USA

S. Kumar (ed.), *An Introduction to Bioanalysis of Biopharmaceuticals*, AAPS Advances in the Pharmaceutical Sciences Series 57,
https://doi.org/10.1007/978-3-030-97193-9_1

1.1 Biopharmaceutical Diversity

This book uses *biotherapeutics* or *biopharmaceuticals* as a broader term interchangeably to cover the diversity of therapeutic modalities—ranging from large-molecule (LM) therapeutics to the emerging modalities (e.g., gene therapy, cell therapy, oligonucleotide therapy) that do not necessarily fall within the conventional categories of LM and small-molecule (SM) therapeutics.

Figure 1.1 illustrates a range of therapeutic modalities that have been explored for the design of the current and future biopharmaceuticals. This book presents the fundamental concepts of bioanalytical characterization for a broad range of biotherapeutics such as monoclonal antibodies, antibody-drug conjugates, bispecific antibodies, fusion proteins, gene therapy, cell therapy, oligonucleotides, and peptides. Small-molecule drugs are out of the scope of this book.

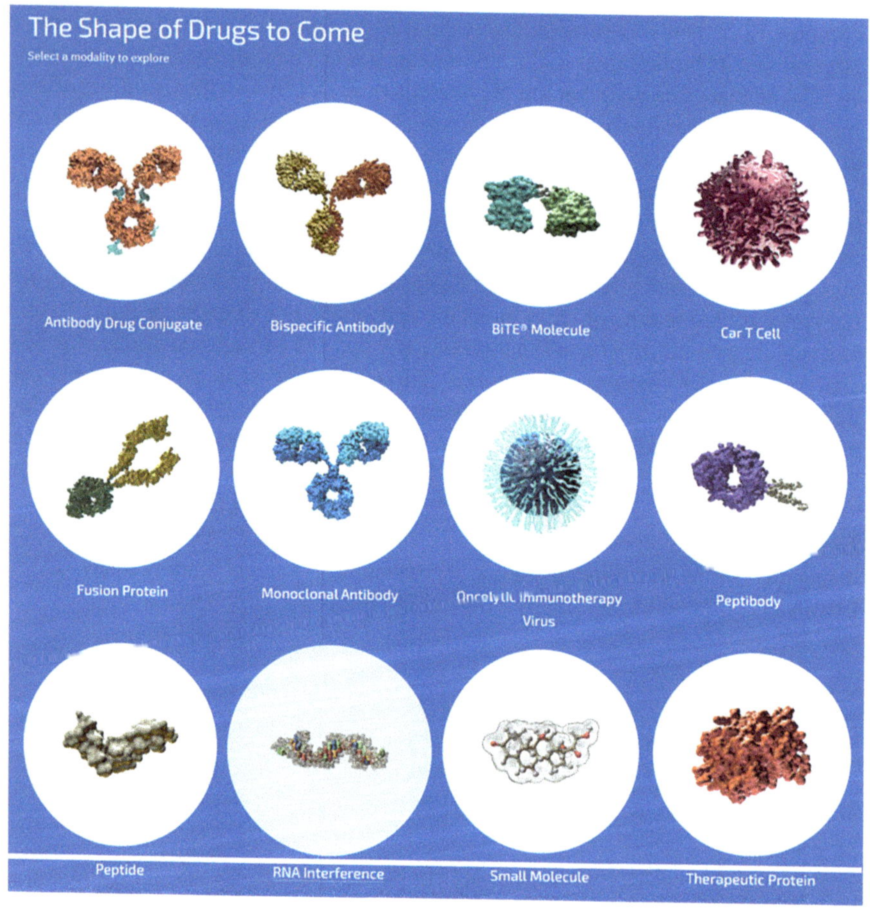

Fig. 1.1 Therapeutic modalities for the design of the current and future biopharmaceuticals. (Source: https://www.amgenscience.com/features/the-shape-of-drugs-to-come/)

1.2 Small-Molecule Vs Large-Molecule Biotherapeutics Bioanalysis

Therapeutic drugs are divided into two categories based on their molecular weight: small-molecule (typically <5 k) and large-molecule (>5 K). Small-molecule therapeutics have been the mainstay of the pharmaceutical industry for decades. These typically include synthetic and low molecular weight small organic molecules, oligonucleotides, and small peptides. The bioanalytical characterization for these therapeutics is well established. The analyte(s) is separated out of the matrix via precipitation and is quantitated by chromatographic technologies coupled with mass spectrometry detection, such as, gas chromatography-mass spectrometry (GC-MS) and liquid chromatography-mass spectrometry (LC-MS). Since low molecular weight molecules are highly homogeneous with unique characteristics, the high-throughput assay designs for their bioanalysis are more amenable.

Large-molecule biotherapeutics on the other hand are derived from biological origin. They encompass a range of modalities such as recombinant proteins, large peptides, antibodies (mono-, bi-, multi-specific), antibody-drug conjugates, fusion proteins, etc. Due to their large molecular size, complex molecular structures, and heterogeneous nature, these molecules are often difficult to fully characterize. Additionally, these molecules could be seen as foreign materials by the immune system and could trigger unwanted immune response. The safety and efficacy of LM biotherapeutic thus requires monitoring of both LM biotherapeutic and the immunogenicity response against LM.

Ligand binding assays (LBAs) have been the primary tools for LM bioanalysis. With the advancement in mass spectrometry technologies, LC-MS/MS-based methods are now increasingly employed to support LM bioanalysis, particularly hybrid LC-MS/MS that combines the high selectivity of LBA and sensitivity of an LC-MS/MS method.

Emerging biotherapeutic modalities such as oligonucleotide, gene, and cell therapy drug candidates that do not necessarily fall within either of the conventional LM and SM therapeutic categories employ a combination of LBA and LC-MS/MS approaches in addition to other modality-specific bioanalytical approaches such as flow cytometry, DNA/RNA quantitation, etc. (Stevenson et.al., 2018). Practical bioanalytical considerations for these biotherapeutics are described in the book.

1.3 Biotherapeutics Drug Disposition

"The dose makes the poison," or in Latin, *Sola dosis facit venenum.* A quote by Paracelsus defines the challenges and intricacies of drug development process especially the challenges when it comes to drug design and its disposition in the body. A molecule can turn into a therapeutic drug or a poison depending on its dosing level and toxic effects.

The safety and efficacy of a therapeutic drug is driven by its pharmacokinetics (PK) and pharmacodynamics (PD) disposition. While PK helps establish the absorption, distribution, metabolism, and excretion (ADME) properties of the therapeutic drug, PD helps determine the pharmacological response of the therapeutic drug (Tibbitts et. al., 2016).

1.3.1 Absorption/Distribution

The disposition of LM biotherapeutics in the body is more complex than SM drugs. Large-molecule therapeutic drug properties such as molecular size, hydrophilicity, and gastric degradation dictates their delivery routes, tissue distribution, metabolism, and excretion. Unlike their SM counterparts, LM biotherapeutics are administered by parenteral routes due to their lack of stability in the gut. The common routes for LM biotherapeutic delivery are intravenous (IV), subcutaneous (SC), or intramuscular (IM).

1.3.2 Metabolism/Elimination

Large-molecule biotherapeutic is absorbed in the cells via receptor-mediated transporters, endocytosis, and/or pinocytosis. The removal of LM biotherapeutic from the circulation is accomplished by target-specific and non-specific pathways. These include target-mediated clearance, Fc gamma-mediated clearance, endocytosis/pinocytosis, degradation by enzymes, and formation of immune complexes that get cleared out by Fc-receptor or complement-mediated clearance pathways. Since the non-specific pathways are not saturated, these clearance pathways contribute to linear PK of LM. The target-mediated clearance, also known as target-mediated drug disposition (TMDD), on the other hand, lends non-linear elimination kinetics for LM biotherapeutics. Once LM biotherapeutic binds to the target, it gets internalized and degraded by lysosomes. As a result, at lower doses, the LM drug clearance remains non-linear. But once the target is saturated at higher doses, the LM clearance becomes linear.

1.3.3 Excretion

The excretion of SM therapeutic is mainly accomplished via renal excretion (kidneys). A low level of SM is also excreted via bile and plasma protein binding. The excretion of LM biotherapeutic from the kidneys is blocked by the charge, structure, and size of the LM drug. The bile pathway to excretion is accomplished by the degradation of the LM drug in the liver and then secretion into the bile. Plasma protein binding of the therapeutic drug dictates the tissue biodistribution of the dosed drug which in turn leads to the degradation/excretion at the tissue level.

1.4 Immunogenicity

Large-molecule biotherapeutics have the potential to elicit humoral and cellular immune response upon administration. The humoral response includes formation of anti-drug antibody (ADA) that can influence the disposition of LM biotherapeutics. The FDA in 2019 released guidance "Immunogenicity Testing of Therapeutic Protein Products—Developing and Validating Assays for Anti-Drug Antibody Detection" which represents current FDA position on developing and validating ADA assays.

Unwanted immune responses to biotherapeutics may pose problems for both patient safety and product efficacy. Large-molecule biotherapeutics have been reported—to elicit immune-mediated safety adverse events, such as anaphylaxis, cytokine release syndrome, and auto-immune disease due to cross-reactive neutralization of non-redundant endogenous proteins, and to reduce therapeutic efficacy by neutralizing the pharmacological activity of the biotherapeutic. Both neutralizing and non-neutralizing ADAs could cause poor efficacy of the LM biotherapeutic and thus need to be thoroughly evaluated. While, non-neutralizing ADA could affect the efficacy of the LM biotherapeutic by impacting its clearance and PK/PD disposition. The neutralizing ADA could reduce the overall efficacy of the LM biotherapeutic by interfering with its biological activity.

Many factors may influence the immunogenicity of biotherapeutics (Zhou and Qiu 2019). These include patient-related, product-related and treatment-related factors. Patient-related factors that might predispose the patient to an immune response include their underlying disease, genetic background, and immune status. Product-related factors also influence the likelihood of an unwanted immune response, e.g., source of protein, manufacturing process (impurity profile, contaminants), formulation and stability characteristics (degradation products, aggregates) of the therapeutic protein. Treatment-related factors include route of administration, therapeutic dose, dosing interval, and duration of treatment. All of these factors are taken into consideration for the LM immunogenicity risk assessment. Depending on the risk assessment, appropriate immunogenicity testing strategy is adopted to measure an immune response against the LM biotherapeutic.

The immunogenicity strategy for various biotherapeutic modalities is covered in great length in the subsequent chapters.

1.5 Bioanalytical Strategies for Biotherapeutics

The bioanalytical strategies used to quantitate biotherapeutics depend on the stage of the drug development process. During early discovery stages, bioanalytical methods are needed to support screening and ranking of multiple biotherapeutic drug candidates. Typically for LBA, during early discovery stages, biotherapeutic-specific critical reagents (such as anti-idiotypic monoclonal antibodies) are not available. Thus, generic reagents such as antibodies directed against the whole human IgG framework or against the (Fab')2 region, or against the Fc region, or against the light chain (LC), or against the heavy plus light (H + L) chain regions, etc. are often employed for early discovery LBA support. The generic reagents are commercially available and can be used to build a fit-for-purpose off-the shelf assay that can be easily adapted from one biotherapeutic to another to provide quick turnaround of the discovery PK data. In some cases, the recombinant target proteins may be used as a critical reagent to develop a surrogate biotherapeutic-specific LBA.

Once the lead candidate(s) is identified, the bioanalytical focus shifts to provide support for the preclinical development of the biotherapeutic. Typically, at this stage, biotherapeutic-specific critical reagent generation is initiated. While the generic LBA may continue to support preclinical development including PK/PD, efficacy, and dose-range finding non-clinical toxicity studies, specific LBA are typically used for late-stage preclinical (pivotal IND-enabling non-clinical toxicity studies) and clinical development. The biotherapeutic-specific critical reagents enable assay development and validation in accordance with regulatory guidelines. Per regulatory guidance, the specificity and the selectivity parameters of the assay are rigorously tested to minimize undesired effects caused by the presence of the biological matrix. The assay should be sensitive enough to measure low concentrations of biotherapeutic to enable exposure-response correlation for safety and efficacy.

In addition to the wet-lab experiments, the validated assays need to adhere to GLP documentation practices. The bioanalytical method and validation plans and reports are appropriately documented as defined by Standard Operating Procedures (SOPs) and GLP guidelines.

The PK/PD and exposure data from efficacy and toxicity studies in animals are used to predict safe starting dose in humans. The clinical development of biotherapeutics involves monitoring the systemic exposure and safety of biotherapeutic drug in humans during Phase 0/1 clinical trials. The dose level is modulated based on the safety profile from the first dosing regimen during Phase 2 and 3. The objective of Phase 2 is to determine the efficacy of the drug in a smaller cohort of patients. The safety and efficacy are evaluated with a larger patient population during Phase 3 and sometimes Phase 4 trials. These trials also help determine if the tested drug is better at curing the disease or provides better quality of life especially for the oncology treatments.

1.5.1 Ligand Binding Assays for Biotherapeutics

Ligand binding assays are based on interaction of biotherapeutic (protein, antibody, or a receptor) with its ligand. Thus, the LBA assays aredeveloped taking into consideration the intended purpose of the assay, the target biology, the mechanism of action, and the physicochemical properties of the biotherapeutic modality.

The LBA platforms range from enzyme-linked immunoassay (ELISA), radioimmunoassay (RIA), to electrochemiluminescence (ECL) and fluorescence (Gyros, Simoa, Quanterix)-based detection of the analyte of interest. While the ECL and fluorescence detection-based LBA platform has become the workhorse for LM biotherapeutics, ELISA and RIA are still being used for niche applications.

The LBA format typically involves a sandwich assay utilizing two antibodies, or combination of recombinant target protein and antibody that bind different epitopes on the biotherapeutic analyte. One antibody of the pair is immobilized to the microtiter plate surface (capture), and the other is conjugated with a detection tag to elicit a color reaction via HRP/AP/B-galactosidase reactions (ELISA), radioactive molecule such as iodine 125 (RIA), an electrochemiluminescence reaction via a ruthenium chelate (ECL), or a fluorescence signal via a probe.

Immunoassay-based approaches have inherent issues such as indirect detection of the analyte of interest, dependence on the quality of critical reagents, and analyte detection directly in the biological matrix. The nonlinearity of the analyte measurements (also referred as Hook effect), performance of reference standards, and capture/detection reagents over assay duration and matrix interference are direct results of these inherent assay complexities. The assay steps such as incubation times, temperature, and light exposure (especially for ECL-based assays) could also turn out to be the likely cause of non-compliance of the assays and thus are critical to the success of the overall LBA performance.

1.5.2 Detection Platforms

The basics of all the instrumentation discussed here rely on an immune complex formation followed by analyte detection in the multiple formats. The assay steps vary from simple ELISA type to a little more involved such as in the PCR amplification. ELISA-based immunoassays don't need a specific instrumentation as the output is read on a spectrophotometer. ECL-based detection platform from Mesoscale discovery supports 96 and 384 well formats for both single and multiplex assays in high-throughput manner. Singulex Erenna instrument is based on fluorescence detection at single-molecule level termed SMC or single-molecule counting. It is a bead/plate-based sandwich assay in 96 and 384 well format that requires a biotinylated capture and a fluorescent labeled detection molecule. Single bead array (Simoa) technology is similar to the Singulex in immunocapture steps but differs slightly in the detection and signal amplification steps to gain in the assay sensitivity. The capture/target is coated onto the plates or magnetic beads surface and is used to capture analyte on a plate, or a disc array followed by addition of a fluorescence producing biotinylated antibody. The microwells on the arrays are designed to capture one immune complexed bead only, and thus the generated fluorescence at bead level is recorded by CCD camera/detector. Immuno-PCR (Imperacer) takes advantage of specificity of capture antibody to target (immune complex) and detection antibody with DNA conjugate in a plate format. The conjugate allows for qPCR signal amplification to yield very high sensitivities. The Immuno-PCR technology is very attractive for immunoassays requiring high sensitivities with sample volume limitations as is the case with nonclinical studies in small animal models.

Flow cytometry is a microfluidics-based platform that analyzes and measures the particles by illuminating them with a laser source. Fluorescent particles are differentiated based on the size and the complexity associated with it at the granular level.

The scope of the method and availability of assay development tools and components must be well defined for a successful assay development. The availability of well-qualified and validated critical reagents, relevant matrices, and data analysis tools is central to any assay development efforts. The choice of monoclonal vs polyclonal antibody is dictated by the scope of the assay, but polyclonal are less preferred due to lower affinity and animal-to-animal variability of generated polyclonal antibodies. Monoclonal antibodies normally have higher analyte affinity that translates into higher specificity and robustness in the assay.

A successful assay should adhere to the regulatory guidelines for bioanalytical validation (Booth 2019; Corr, 1922; Food and Drug Administration, 2018). The assay validation documentation should include information about the reference standards and critical reagents, optimization of calibration standards and limit of quantitation, minimal required dilution determination, specificity, selectivity, dilutional linearity, prozone or hook effect, and analyte stability. Any changes in the assay performance and parameters should be supported by strong scientific rationale to justify the change(s). Although assay development doesn't require extensive planning and record keeping, it is advised to generate assay development summary table and report as it may aid in the assay troubleshooting and/or assay investigation during validation and/or sample testing.

A partial assay validation is warranted when slight modifications to the assay are made. The assay transfer to another site/laboratory/CRO requires a cross validation of the assay.

A fully characterized reference standard (similar to the analyte being measured) with a Certificate of Analysis (CoA) is required to ensure that the quality of the reference standard is maintained throughout the assay validation and sample testing.

1.5.3 Direct and Indirect Measurement of Biotherapeutics Using LC-MS

Ligand binding assays (LBAs) such as enzyme-linked immunosorbent assay (ELISA) are essential to measure large molecules including proteins, peptides, antibodies (mono-, bi-, tri-specific), antibody-drug conjugates, fusion proteins, oligonucleotides, and gene therapy candidates. Having reliable method platforms that can quantify these molecules in biological fluids is critical for drug development for efficacy, toxicity, and immunogenic evaluations (DelGuidice et al. 2020). LBA's indirect drug measurement at the binding site may not be detected due to functionality changes such as a glycan attachment (Cymer et. al. 2018, Higel et. al., 2016, Liu et. al. 2015, Zhou and Qiu 2019). As a complement to these assays, liquid chromatography coupled with mass spectrometry (LC-MS) has been increasingly used for improving selectivity and linear dynamic range. LBAs also lack specificity and are incapable of structural characterization; thus, LC-MS is especially important for applications such as antibody-drug conjugates (Kang et al. 2020). LC-MS method development time can be much shorter, especially if critical reagents are not required during early drug discovery programs. In order to accomplish these advances, one approach to large-molecule LC-MS/MS analysis involves indirect measurement (i.e., bottom-up) using unique peptides known as signature peptides as a surrogate for the biologic of interest. Signature peptides undergo the same validation rigor to ultimately provide precise and accurate pharmacokinetic data. Direct or intact measurements (i.e., top-down) of biotherapeutics using LC-MS, albeit with high-resolution accurate mass (HRAM) coupled with liquid chromatography, are another approach continuously evolving. Sample processing, internal standardization, separation, detection, and data analysis are all critical aspects to consider for maintaining the integrity of a signature peptide or intact biotherapeutic quantification. In this section, intact (top-down) and signature peptide (bottom analysis) will be discussed. A summary of target protein quantification is represented in Fig. 1.2.

1.5.3.1 Signature Peptides

Indirect or bottom-up measurement of proteins require digestion into small peptides that are analyzed as surrogates. The first step is to choose signature peptides for quantitative analysis. Data-dependent acquisition, database review, and predictive methods are essential to obtaining appropriate signature peptides. Identifying unique peptides that can represent the protein(s) of interest requires initial primary structure in silico predictions using software-based platforms such as Skyline. Proteome Discoverer can be used to obtain spectral files and rank peptides. Obtaining signature peptides in an iterative process involving in silico and experimental selection, ultimately to reach a multiple reaction monitoring (MRM)-based method for quantitative analysis (Fu et al. 2016). Initially, target peptides will be characterized with primary structure information. Inclusion and exclusion criteria exist to ensure

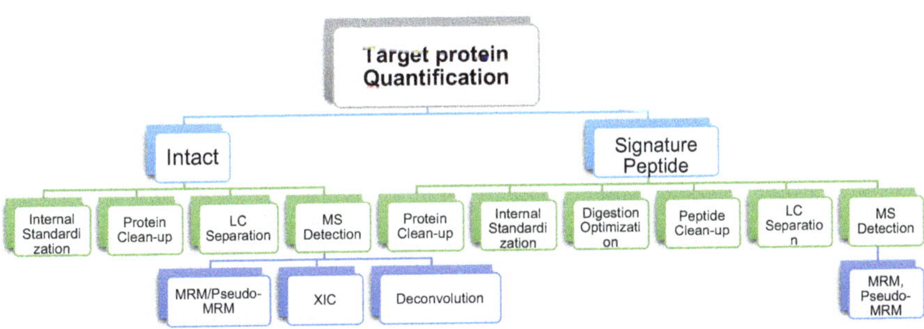

Fig. 1.2 Overall strategies for target protein quantification using intact or signature peptide approaches

the protein surrogate will be unique for quantitative analysis. These include the following: (1) Apply sequence to a Basic Alignment Search Tool (BLAST) analysis; (2) input an enzyme, most often used would be trypsin; (3) select peptides with 5–21 amino acids; (4) determine any peptides for post-translational modifications (PTMs) and isoforms; and (5) exclude amino acid residues like cysteine and methionine. Once theoretical peptides have been established, empirical and experimental steps are taken to ensure optimal signature peptides. Initially, (1) trypsin digestion will be optimized, (2) peptide clean-up is done including chromatography, and (3) which signature peptides that are most abundant are determined along with selective SRM transitions. Pre-digestion steps to ensure selectivity, efficient digestion, and recovery are also an integral part of obtaining a quantitative method. A summary of empirical and experimental optimization is depicted in Fig. 1.2. Transitioning into the quantitative stage, several components of method development and validation are necessary. Having synthetic peptides in both natural and isotopically labelled forms is key to optimization of SRM parameters (Fu et al. 2016; Picotti and Aebersold 2012). Various vendors (i.e., Elim Biopharmaceuticals, Thermo Fisher Scientific, Biomatik) offer services to provide custom synthetic peptides. A certificate of analysis containing MS spectra, HPLC chromatograms, and, most critical, high purity are necessary for ultimately reaching a validated method. These synthetic peptides will be used to optimize chromatography and SRM parameters and evaluate preliminary validation studies (i.e., selectivity, recovery, reproducibility) (Fig. 1.3).

1.5.3.2 Top-Down or Intact Measurement

Challenges still exist with the signature peptide approach although with the exception of potentially highly sensitive MS-based methods. Starting with the obvious, the approach does not detect the intact protein. Structural and biotransformation information will be lost during enzymatic digestion. Opposite to the bottom-up/signature peptide approach, whole-molecule quantitation delivers structural and high-level sequence information. Unlike the commonly used triple quadrupoles for bottom-up analysis of signature peptides, intact analysis of proteins requires high resolution mass spectrometry (HRMS). Higher resolving power outdoes triple-stage quadrupoles such as an Orbitrap (i.e., linear, Q-Orbitrap), Time-Of-Flight (i.e., TOF, QTOF, TOF-TOF), or Fourier transform ion cyclotron resonance (FTICR) mass spectrometers (Kellie et al. 2016; Ramagiri and Garofolo 2012). The selectivity of intact structural preservation is beneficial for any post-translational modifications while quantitative analysis can be achieved, albeit with lower sensitivity. Another advantage with intact analysis of proteins is the removal of the digestion step and preceding processes (i.e., denaturation). This allows a decrease in sample processing time, incomplete digestion, and potentially lost peptide information. Internal standardization is essential for all MS-based assays, and obtaining fully intact stable isotope labeled (SIL) internal standard proteins is difficult to produce, time-consuming, and costly (DelGuidice et al. 2020; Kang et al. 2020). Alternative

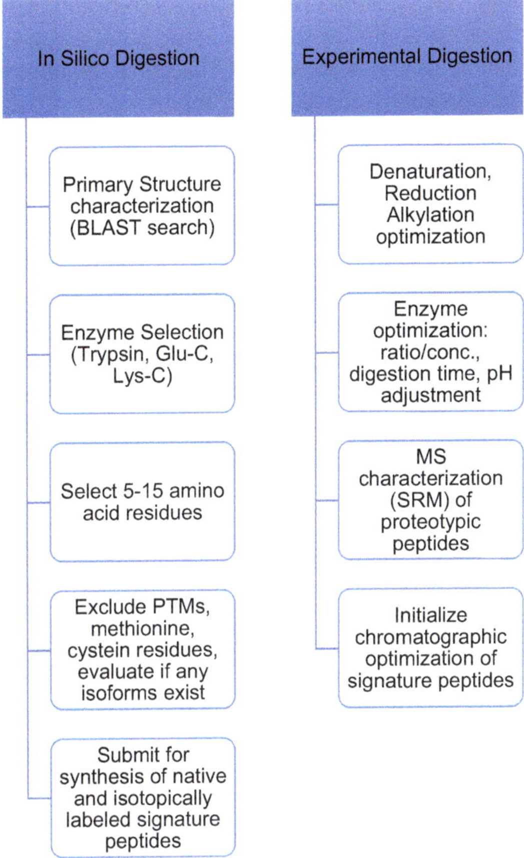

Fig. 1.3 Empirical and experimental optimization to achieve a signature peptide

strategies that are less specific such as SILu™Mab are available for monoclonal IgG antibodies (Li et al. 2017) and can be applied to generic monoclonal antibody studies. Protein analogs have also been used for internal standards; however, using immunocapture to clean up the protein of interest would hinder the state at which an analog could be added. In addition to sample preparation, which will be discussed further in the next section, data processing is expansive for intact analysis. The benefits of MRM transitions for absolute quantification are unmatched, yet not feasible for intact analysis due to variations in charge state distribution, fragmentation reproducibility, and efficiency. There is a balance in intact molecules with collision-induced dissociation and optimal collision energy to achieve effective fragmentations (i.e., product ions). Other studies have attempted pseudo-MRM, using identical precursors and product ions (Hammond et al. 2016). Intact analysis may not be suitable in this pseudo-MRM state, but the molecule of interest, matrix of choice, and desired detection limits play a decisive role. These approaches that lack reliability have led to the use of HRMS, owing to its resolving power (up to 1 M FWHM) and ability to characterize molecules with full-scan, extracted ion chromatograms (XIC), summed ions (XICs), and zero charge state deconvolution. HRMS systems such as Q-orbitraps have the ability to scan the entire mass range (m/z) yet supply the selectivity to distinguish between similar molecules. Summing of extracted ion chromatograms has shown to be beneficial to improving sensitivity and is software dependent (Kang et al. 2020; Qiu et al. 2018; Ramagiri and Garofolo 2012; Tassi et al. 2018). Deconvolution to obtain zero charge state mass is another technique for intact data processing. Software will obtain the entire mass spectrum (isotopes and charge state distribution) through deconvolution algorithms. A respective spectrum after deconvolution will reveal a comparative intensity to mass spectrum, obtained from the previous charge states in the original spectra (Sturm et al. 2016; van den Broek and van Dongen 2015; Whitelegge 2013). Advancements in mass analyzers have improved the reality of intact analysis of target proteins in biological fluids; however, sensitivity is still lower (>ng/mL). Overcoming sensitivity challenges in addition to instability (i.e., oxidation, protease activity, or nonspecific binding) is a part of what analysts will face with target protein analysis.

1.5.3.3 Sample Preparation for Signature Peptides (Bottom-Up Quantification)

During protein quantification, steps are necessary to enrich or purify the protein as well as in the peptide state depending on which approach is used. Protein purification can consist of simply a buffer dilution, partial precipitation, abundant protein removal, or most preferably an immunocapture prior to enzymatic digestion for signature peptide or bottom-up analysis (Chappell et al. 2014; Halquist and Karnes 2011; Zhang and Jian 2014). Dilution solvents consist of a buffer such as ammonium bicarbonate, either for immunocapture or digestion accommodation. Prior to enzymatic digestion, the native protein may undergo partial precipitation with a pH adjustment outside of its isoelectric point as one approach. If quantification is in plasma, abundant protein removal is a secondary option, where the top 20 most abundant proteins represent 99% of total protein (Zhang and Jian 2014). Removal or depletion of proteins like albumin (30–50 mg/mL) can alleviate a significant portion of protein content; however, some globular proteins may be nonspecifically bound to albumin. Depletion kits can be purchased to remove the top concentrated proteins and remove up to 85% (Echan et al. 2005). As demonstrated in previous studies (L. Anderson and Hunter 2006; Polaskova et al. 2010), between six and 11 proteins have been removed. Briefly, depletion kits of affinity gel-based systems and columns are employed to specifically remove up to the top 20 proteins in plasma. Normally proteins will bind to these columns, and eluents such as urea (i.e., 6 M) and buffer (CHAPs or Tris) may be combined with the idea of recovering low abundant proteins of interest for further preparation (i.e., digestion). Immunocapture or immunoaffinity is the most effective enrichment of target proteins.

Following protein enrichment or purification, enzymatic digestion processes transpire. In order to achieve maximum digestion efficiency, denaturation, reduction, and alkylation will precede normally enzymatic digestion, mostly using trypsin. Denaturation allows for the opening of the tertiary and quaternary structure by breaking of weaker bonds (i.e., hydrogen) thus allowing better exposure to digestible peptide bonds. Heat, use of chaotropic agents (i.e., urea), acids (i.e., formic acid), surfactants (i.e., RapiGest), and some organic solvents all can induce denaturation (Bennion and Daggett 2003). Following denaturation, reduction and alkylation will take place prior to digestion. Reduction will break disulfide bonds on cysteine residues, commonly by the addition of dithiothreitol (Müller and Winter 2017). After reduction, alkylation is performed with reagents like iodoacetamide, to prevent the free sulfhydryl (SH-) groups from forming new disulfide bonds. Denaturation, reduction, and alkylation have a synergistic effect on enzymatic digestion. Following digestion, peptide cleanup may be necessary to reduce matrix effects and improve recovery. Cleanup may be similar to small-molecule analysis using dilution, protein precipitation, solid-phase extraction, or even employing immunoaffinity techniques (i.e., SISCAPA) (N. L. Anderson et al. 2004). Top-down or intact sample preparation is approached either in native or under denatured conditions. Following denaturing, an unfolded protein exposes more amino acid sites for protonation, thus enhancing charge state. This will be critical for analyzers with mass-to-charge ratio (m/z) limitations and defined resolving power. Most sample preparation involves an immunocapture step with a capture reagent such as a monoclonal antibody, anti-idiotypic antibody, Protein A, Protein G, or an anti-human Fc [REF]. These critical reagents are normally immobilized on a support system such as a magnetic bead coated with streptavidin in combination with a biotinylated capture antibody [REF]. The antigen antibody complex will be washed thoroughly and then may or may not be separated by lowering the pH or changing the ionic state prior to mass detection. Optimizing conditions is critical for improving yield which includes order of binding, antibody concentration, elution, and incubation [REF]. Reagent compatibility should also be adapted to LC-MS conditions, where mobile-phase composition with the sample is injected. Additional sample preparations such as filtration, molecular weight cutoff, or solid-phase extraction may also be incorporated prior to LC-MS [REF]. Intact analysis requires preservation of the native conditions; therefore, maintaining physiological state (i.e., pH 7.4) is necessary to measure intact proteins. Buffers such as ammonium acetate or bicarbonate, dilutions with reagent water, and desalting are all considered to obtain these conditions [REF]. Deglycosylation may also be needed using an enzymatic digestion with PNGase F or EndoS to remove N-linked oligosaccharides [REF]. Following sample cleanup, intact or peptide separation is generally reversed-phase (RPLC) or hydrophobic interaction (HILIC) chromatography. Stationary phases may not be as hydrophobic as a traditional C18 for optimal separation; therefore, C4, C8, or phenyl should be evaluated. For intact analysis, ion exchange chromatography (IEC) is also employed, where a pH gradient is used alongside a stationary phase such as a PolyCAT A, with mobile phases such as diethanolamine buffer [REF]. Two-dimensional chromatography (2DLC) may also be incorporated with RPLC and size exclusion (SEC). SEC could first separate molecules by size followed by RPLC or IEC. When choosing stationary phases, pore size is crucial for appropriate separation especially for intact analysis. Therefore, pore sizes ranging from 200 to 4000 should be evaluated. Other considerations for chromatography include column temperature, flow (i.e., microflow or conventional), mobile-phase additives, and strength (i.e., formic acid, trifluoracetic acid, acetic acid).

1.6 Biotherapeutics Bioanalytical Challenges and Future Directions

Some of the major challenges associated with bioanalysis include achieving desired sensitivity, selectivity, and specificity in the presence of biological matrix (plasma, serum, urine, CNS, etc.). Additionally, biotherapeutic properties such as the analyte structure and stability in matrices could also present challenges in bioanalysis. For instance, LBA may not quantify truncated or PTM modified biotherapeutic analyte(s) due to biotransformations (Cymer et. al. 2018, Higel et. al., 2016, Liu et. al. 2015, Zhou and Qiu 2019) and thus an orthogonal confirmatory approach may be needed to gain full understanding of the biotherapeutic disposition. The combination of LBA with LC-MS/MS approach leverages advantages of both LBA and LC-MS/MS analytical platforms for biotherapeutics quantitation (Amaravadi et. al., 2015, Andrews et.al., 2015, Fischer et al., 2015, Zhang et. al., 2017).

The current practice of leveraging combination of bioanalysis in tissue samples also poses challenges in gaining comprehensive understanding of the tissue biodistribution of biotherapeutics. The disadvantage of the loss of the spatial information/localization of the biotherapeutic due to sample preparation for LBA analysis could be overcome by leveraging orthogonal techniques such as LC-MS and imaging mass spectrometry/cytometry.

The adoption of singlicate analysis is being considered in the bioanalytical field. This effort could lead to increased efficiency due to better optimization of plate area, sample amounts, and resourcing of laboratory scientists.

The unwanted consequences of immune responses to biotherapeutic products can range from no apparent effect to serious adverse events, including life-threatening complications such as anaphylaxis, neutralization of the effectiveness of life-saving or highly effective therapies, or neutralization of endogenous proteins with non-redundant functions. Because T-cell epitopes are necessary for a robust humoral response, accurate T-cell epitope predictions will correlate to the actual response in vivo. The implementation of regular protocols for screening therapeutic proteins in vitro studies may allow researchers to avoid the development of ADA and may also reduce the costs of recombinant protein drug development by eliminating candidates that are determined to be too immunogenic. Current models used to predict immunogenicity have enabled the pharmaceutical industry to identify factors that can be controlled during drug development and thus guide protein design and candidate selection at early stages of drug development.

Disclaimer Any opinions or forward-looking statements expressed are those of the authors and may not reflect views held by their employers (Janssen Therapeutics for Sanjeev Bhardwaj, Spark Therapeutics for Inderpal Singh, and Virginia Commonwealth University for Matt Halquist).

References

Amaravadi L, Song A, Myler H, Thway T, Kirshner S, Devanarayan V, et al. 2015 White Paper on recent issues in bioanalysis: focus on new technologies and biomarkers (part 3-LBA, biomarkers and immunogenicity). Bioanalysis. 2015;7:3107–24.

Anderson L, Hunter CL. Quantitative mass spectrometric multiple reaction monitoring assays for major plasma proteins. Mol Cell Proteomics. 2006;5(4):573–88. https://doi.org/10.1074/mcp.M500331-MCP200.

Anderson NL, Anderson NG, Haines LR, Hardie DB, Olafson RW, Pearson TW. Mass spectrometric quantitation of peptides and proteins using stable isotope standards and capture by anti-peptide antibodies (SISCAPA). J Proteome Res. 2004;3(2):235–44. https://doi.org/10.1021/pr034086h.

Andrews L, Ralston S, Blomme E, Barnhart K. A snapshot of biologic drug development: challenges and opportunities. Hum Exp Toxicol. 2015;34:1279–85.

Bennion BJ, Daggett V. The molecular basis for the chemical denaturation of proteins by urea. Proc Natl Acad Sci. 2003;100(9):5142–7. https://doi.org/10.1073/PNAS.0930122100.

Booth B. ICH M10 Bioanalytical Method validation guidelines, 2019.

Chappell DL, Lassman ME, McAvoy T, Lin M, Spellman DS, Laterza OF. Quantitation of human peptides and proteins via MS: review of analytically validated assays. Bioanalysis. 2014. Future Science Ltd;6(13):1843–57. https://doi.org/10.4155/bio.14.145.

Corr, Guideline on bioanalytical method validation, EMEA, 1922;44:1–23.

Cymer F, Beck H, Rohde A, Reusch D. Therapeutic monoclonal antibody N-glycosylation—structure, function and therapeutic potential. Biologicals. 2018;52:1–11. https://doi.org/10.1016/j.biologicals.2017.11.001. [Internet]. Elsevier

DelGuidice CE, Ismaiel OA, Mylott WR Jr, Halquist MS. Quantitative bioanalysis of intact large molecules using mass spectrometry. J Appl Bioanalysis. 2020;6(1):52–64. https://doi.org/10.17145/jab.20.006.

Echan LA, Tang H-Y, Ali-Khan N, Lee K, Speicher DW. Depletion of multiple high-abundance proteins improves protein profiling capacities of human serum and plasma. Proteomics. 2005;5(13):3292–303. https://doi.org/10.1002/pmic.200401228.

Fischer SK, Joyce A, Spengler M, Yang TY, Zhuang Y, Fjording MS, et al. Emerging technologies to increase ligand binding assay sensitivity. AAPS J. 2015;17:93–101.

Food and Drug Administration, Guidance for Industry Bioanalytical Method Validation Guidance for Industry Bioanalytical Method Validation. Draft Version Non-Binding Recomm by FDA. 2018; Rev. 1.

Food and Drug Administration. Guidance Document. Immunogenicity assessment for the therapeutic protein products—Developing and validating assays for anti-drug antibody detection. 2019 [Docket Number: FDA-2009-D-0539]. https://www.fda.gov/regulatory-information/search-fda-guidance-documents/immunogenicity-testing-therapeutic-protein-products-developing-and-validating-assays-anti-drug

Fu Q, Grote E, Zhu J, Jelinek C, Köttgen A, Coresh J, Van Eyk JE. An empirical approach to signature peptide choice for selected reaction monitoring: quantification of uromodulin in urine. Clin Chem. 2016;62(1):198–207. https://doi.org/10.1373/clinchem.2015.242495.

Guideline on bioanalytical method validation, EMEA, 1922;44:1–23.

Halquist MS, Karnes HT. Quantification of Alefacept, an immunosuppressive fusion protein in human plasma using a protein analogue internal standard, trypsin cleaved signature peptides and liquid chromatography tandem mass spectrometry. J Chromatogr B Anal Technol Biomed Life Sci. 2011;879:11–2. https://doi.org/10.1016/j.jchromb.2011.02.034.

Hammond TG, Moes S, Youhanna S, Jennings P, Devuyst O, Odermatt A, Jenö P. Development and characterization of a pseudo multiple reaction monitoring method for the quantification of human uromodulin in urine. Bioanalysis. 2016;8(12):1279–96. https://doi.org/10.4155/bio-2016-0055.

Higel F, Seidl A, Sörgel F, Friess W. N-glycosylation heterogeneity and the influence on structure, function and pharmacokinetics of monoclonal antibodies and fc fusion proteins. Eur J Pharm Biopharm. 2016;100:94–100. [Internet]. Elsevier B.V. https://doi.org/10.1016/j.ejpb.2016.01.005.

Kang L, Weng N, Jian W. LC–MS bioanalysis of intact proteins and peptides. Biomed Chromatogr. 2020;34:1. https://doi.org/10.1002/bmc.4633.

Kellie JF, Kehler JR, Mencken TJ, Snell RJ, Hottenstein CS. A whole-molecule immunocapture LC-MS approach for the in vivo quantitation of biotherapeutics. Bioanalysis. 2016;8(20):2103–14. https://doi.org/10.4155/bio-2016-0180.

Li W, Lin H, Fu Y, Flarakos J. LC–MS/MS determination of a human mAb drug candidate in rat serum using an isotopically labeled universal mAb internal standard. J Chromatogr B Anal Technol Biomed Life Sci. 2017;1044–1045:166–76. https://doi.org/10.1016/j.jchromb.2016.12.044.

Liu L. Antibody glycosylation and its impact on the pharmacokinetics and pharmacodynamics of monoclonal antibodies and fc-fusion proteins. J Pharm Sci [Internet]. 2015;104:1866–84. Elsevier Masson SAS. https://doi.org/10.1002/jps.24444.

Müller T, Winter D. Systematic evaluation of protein reduction and alkylation reveals massive unspecific side effects by iodine-containing reagents. Mol Cell Proteomics. 2017;16(7):1173–87. https://doi.org/10.1074/mcp.M116.064048.

Picotti P, Aebersold R. Selected reaction monitoring-based proteomics: workflows, potential, pitfalls and future directions. Nat Methods. 2012;9(6):555–66. https://doi.org/10.1038/nmeth.2015.

Polaskova V, Kapur A, Khan A, Molloy MP, Baker MS. High-abundance protein depletion: comparison of methods for human plasma biomarker discovery. Electrophoresis. 2010;31(3):471–82. https://doi.org/10.1002/elps.200900286.

Qiu X, Kang L, Case M, Weng N, Jian W. Quantitation of intact monoclonal antibody in biological samples: comparison of different data processing strategies. Bioanalysis. 2018;10(13):1055–67. https://doi.org/10.4155/bio-2018-0016.

Ramagiri S, Garofolo F. Large molecule bioanalysis using Q-TOF without predigestion and its data processing challenges. Bioanalysis. 2012;4(5):529–40. https://doi.org/10.4155/bio.12.10.

Stevenson L, Richards S, Pillutla R, Torri A, Kamerud J, Mehta D, et al. 2018 white paper on recent issues in bioanalysis: focus on flow cytometry, gene therapy, cut points and key clarifications on BAV (part 3—LBA/cell-based assays: immunogenicity, biomarkers and PK assays). Bioanalysis. 2018;10:1973–2001.

Sturm RM, Jones BR, Mulvana DE, Lowes S. HRMS using a Q-Exactive series mass spectrometer for regulated quantitative bioanalysis: how, when, and why to implement. Bioanalysis. 2016;8(16):1709–21. https://doi.org/10.4155/bio-2016-0079.

Tassi M, De Vos J, Chatterjee S, Sobott F, Bones J, Eeltink S. Advances in native high-performance liquid chromatography and intact mass spectrometry for the characterization of biopharmaceutical products. J Sep Sci. 2018;41(1):125–44. https://doi.org/10.1002/jssc.201700988.

Tibbitts J, Canter D, Graff R, Smith A, Khawli LA. Key factors influencing ADME properties of therapeutic proteins: a need for ADME characterization in drug discovery and development. MAbs. [Internet]. Taylor & Francis. 2016;8:229–45. https://doi.org/10.1080/19420862.2015.1115937.

van den Broek I, van Dongen WD. LC–MS-based quantification of intact proteins: perspective for clinical and bioanalytical applications. Bioanalysis. 2015;7(15):1943–58. https://doi.org/10.4155/bio.15.113.

Whitelegge J. Intact protein mass spectrometry and top-down proteomics. Expert Rev Proteomics. 2013;10(2):127–9. https://doi.org/10.1586/epr.13.10.

Zhang SW, Jian W. Recent advances in absolute quantification of peptides and proteins using LC-MS. Rev Anal Chem. 2014;33(1):31–47. https://doi.org/10.1515/revac-2013-0019.

Zhang YJ, An HJ. Technologies and strategies for bioanalysis of biopharmaceuticals. Bioanalysis. 2017;9:1343–7.

Zhou Q, Qiu H. The mechanistic impact of N-glycosylation on stability, pharmacokinetics, and immunogenicity of therapeutic proteins. J Pharm Sci. 2019;108:1366–77. https://doi.org/10.1016/j.xphs.2018.11.029. [Internet] Elsevier Ltd:

Chapter 2
An Introduction to Bioanalysis of Monoclonal Antibodies

Varun Ramani, Sanjeev Bhardwaj, and Omnia A. Ismaiel

Abstract Therapeutic monoclonal antibodies are one of the largest and fastest growing class of large-molecule biotherapeutics. Accurate and reliable quantification of these biotherapeutics in biological matrix is mandatory for their pharmacokinetic and pharmacodynamic assessments. Bioanalytical assays thus play an integral role in determining safety and efficacy of therapeutic monoclonal antibodies. Both immunoassays and liquid chromatography with tandem mass spectrometry (LC-MS/MS) based bioanalytical assays have been used to follow disposition of these biotherapeutics. This chapter provides an introductory overview of the therapeutic monoclonal antibodies, bioanalytical assays used for their pharmacokinetics and immunogenicity assessments, and regulatory considerations for the bioanalytical assays.

Keywords Monoclonal antibody · Pharmacokinetics · Bioanalytical · Ligand binding assay · LBA-LC-MS/MS · Immunocapture · Surrogate peptides · Internal standard · Immunogenicity · Anti-drug antibody · Screening cut point · Confirmatory/specificity cut point · Titer cut point · Neutralizing antibody

Abbreviations

ADA	Anti-drug antibody
ADC	Antibody drug conjugate

V. Ramani (✉)
Inhibrx, La Jolla, CA, USA

S. Bhardwaj
Janssen Biotherapeutics, Spring House, PA, USA

O. A. Ismaiel
Department of Pharmaceutical Analytical Chemistry, Faculty of Pharmacy, Zagazig University, Ash Sharqiyah, Egypt

S. Kumar (ed.), *An Introduction to Bioanalysis of Biopharmaceuticals*, AAPS Advances in the Pharmaceutical Sciences Series 57,
https://doi.org/10.1007/978-3-030-97193-9_2

ADCC	Antibody-dependent cellular cytotoxicity
Anti-ID Ab	Anti-idiotype antibodies
C1q	Complement 1
CDC	Complement-dependent cytotoxicity
CDR	Complementarity determining region
DTT	Dithiothreitol
EMA	European Medicines Agency
Fab	Fragment antigen binding region of the mAb
Fc	Crystallizable fragment region of mAb
FcRn	Neonatal Fc receptor
FDA	US Food and Drug Administration
IAA	Iodoacetic acid
IAM	Iodoacetamide
IM	Intramuscular
IgG	Immunoglobulin G
IV	Intravenous
LBA	Ligand binding assay
LC-MS/MS or LCMS	Liquid chromatography with tandem mass spectrometry
LLOQ	Lower limit of quantitation
mAb	Monoclonal antibody
PD	Pharmacodynamics
PK	Pharmacokinetics
RES	Reticuloendothelial system
SC	Subcutaneous
scfv	Single-chain variable fragment
SIL-IS	Stable isotopically labeled internal standard
TNF-α	Tumor necrosis factor α
TCEP	Tris (2-carboxyethyl) phosphine hydrochloride
VEGF-A	Vascular endothelial growth factor type A

2.1 Introduction

Monoclonal antibodies (mAbs) are large heterodimeric proteins that are composed of four polypeptide chains of variable sizes. There are five classes of antibodies - IgA, IgD, IgE, IgG, and IgM. Majority of the marketed therapeutic mAbs belong to the IgG class. This chapter focuses on the IgG based therapeutic mAbs. There are four subclasses of IgG that are present in humans—IgG1, IgG2, IgG3, and IgG4. Each IgG contains two identical heavy (H) (~50 kDa) and two identical light (L) chains (~25 kDa) that are held together by disulfide bridges to form a Y-shaped structure. Refer to Fig. 2.1 for the structure of a monoclonal antibody. Each heavy and light chain has a constant (C) and a variable (V) domain, designated as C_H or C_L and V_H or V_L, respectively. The antigen binding fragment (Fab) of monoclonal antibody (mAb) is comprised of a variable region formed by

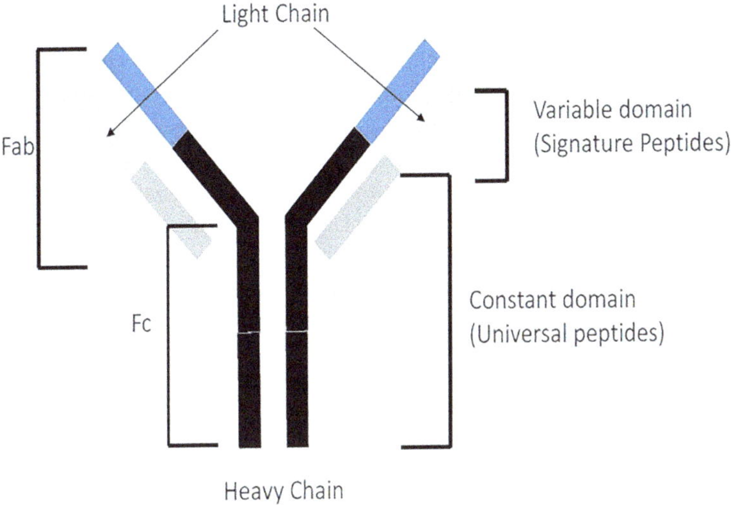

Fig. 2.1 Structure of a monoclonal antibody

variable domains (V_H and V_L-) and a constant region formed by constant domains (C_H and C_L) of heavy and light chains. The Fc or crystallizable fragment region of the antibody consists of two constant domains of the heavy chain.

The Fc region of antibody binds to cell surface receptors such as Fc-gamma (Fcγ), Complement 1 (C1q) protein, and neonatal Fc (FcRn) receptors. The binding of antibody to specific antigens or cell surface receptors can be engineered for the desired pharmacological function(s) leading to disease intervention. The antibody binding to its antigen(s) and/or cell surface receptors drives its pharmacokinetic (PK) and pharmacodynamic (PD) properties. The Fc-dependent pharmacological activity of antibody includes effector functions, such as antibody-dependent cellular cytotoxicity (ADCC), complement-dependent cytotoxicity (CDC), antibody-dependent cellular phagocytosis (ADCP), etc. (Liu 2018; Mahomed et al. 2019).

2.2 Pharmacokinetic/Pharmacodynamic Properties

Monoclonal antibody typically exhibits a biphasic PK profile, consisting of a fast distribution phase followed by a slower elimination phase. The antibody molecular size, polarity/hydrophilicity, and the route of administration affect its distribution (in vasculature and interstitial space), metabolism, and excretion out of the body. Additionally, physicochemical attributes of mAb, such as molecular charge, iso-electric point (pI), hydrophobicity, self-interaction (propensity to aggregate), target binding affinity, off-target binding, pH-dependent FcRn binding affinity, degree and type of glycosylation also drive its PK properties. For instance, while FcRn binding extends half-life of mAb, the target-mediated drug disposition (TMDD) can lead to its increased clearance and shortened half-life (Liu 2018; Mahomed et al. 2019; Mould and Sweeney 2007; Ovacik and Lin 2018).

The common routes of mAb administration are intravenous (IV), subcutaneous (SC), and intramuscular (IM) injection. SC administration, although a preferred route from clinical operations perspective, leads to slow absorption of mAb due to its large size. SC administration thus requires higher doses with efficacious concentration taking hours to days (3–7 days) to achieve 50–90% bioavailability in humans (Datta-Mannan 2019; Keizer et al. 2010; Liu 2018; Mou et al. 2018; Vugmeyster 2012). Additionally, after SC administration, mAb enters circulation through the lymphatic system. As a result, the physiological factors such as blood flow and skin morphology can also affect circulating concentration and bioavailability of mAb.

The degradation and metabolism of mAb to small peptides and amino acids is the major route of its elimination from the body. The other non-specific clearance routes of mAb include Fcγ-mediated clearance, endocytosis, degradation by enzymes, and formation of immune complexes.

The mAb biotherapeutic when administered into clinical and non-clinical subjects can invoke an unwanted immunogenic response (anti-drug antibodies, ADA) against the mAb biotherapeutic. Three major risk factors that drive the immunogenic response against mAb biotherapeutic include patient-related, treatment-related, and mAb drug product-related factors. The patient's genetic background, immune status, and pre-existing immunity could influence the ADA response against mAb biotherapeutic. The treatment related factors, such as route of administration, therapeutic doses, dosing frequency, etc., also influence ADA response. The product stability and manufacturing process-related factors could also contribute to the unfavorable ADA response against the mAb biotherapeutics.

A comprehensive understanding of integrated pharmacology, PK, PK/PD and safety are integral to the development of mAb biotherapeutics.

2.3 Pharmacokinetic Assessment

The PK assay considerations for mAb biotherapeutics are outlined in Chap. 1 The PK assay strategy for mAb biotherapeutic is dependent on the PK question that needs to be addressed at the given stage of drug development. The preclinical PK, PK/PD, safety, and efficacy studies provide insights into the safety and efficacy of mAb biotherapeutic. The dose selection and dosing regimen for efficacy and safety studies are dependent on the PK properties of mAb biotherapeutic.

The accurate quantitation of mAb biotherapeutic is important for determining its PK profile. The bioanalytical assays that quantitate mAb biotherapeutic thus play an important role in the development of mAb biotherapeutics. The robustness and reliability of bioanalytical assays are governed by the critical reagents used in these assays. Both generic anti-human IgG and mAb biotherapeutic-specific reagents are used for mAb quantitation. The bioanalytical assays supporting regulated non-clinical and clinical studies are validated per regulatory guidelines (FDA 2018; M10 2019; European Medicines Agency 2011).

2.3.1 PK Quantitation by Immunoassay

The key for developing successful bioanalytical assay is to understand the nature of the analyte that needs to be quantified by the assay. The bioanalytical assay for mAb quantitation typically involve ligand-binding assay (LBA) platforms. The LBA platforms require either mAb biotherapeutic-specific reagents (such as recombinant target protein, anti-idiotypic Ab, polyclonal Ab, etc.) or generic anti-human IgG reagents. The recombinant target protein or monoclonal antibody is either procured commercially or generated in-house. Polyclonal antibodies, in general are easier to generate than monoclonal antibodies. The need for generation of mAb biotherapeutic-specific reagents add to the assay development timelines and thus are recommended to be explored at the early stages of the assay planning activities. Depending on the PK question to be answered, the target-bound, target-unbound or total (target-bound and target-unbound) antibody assay may be desirable (Lee et al. 2011; Talbot et al. 2015). The presence of soluble or shed target increases the complexity of bioanalytical assays. The development of free and/or total antibody assay depends not only on the specific PK needs of the program but also on the availability of critical reagents.

While, the free mAb biotherapeutic assay is used to determine the target protein coverage required for the desired PD effects. In order to gain a better understanding of the adverse effect at the varying concentrations of the therapeutic protein, a total mAb biotherapeutic assay is normally employed for exploratory toxicity and toxicokinetic (TK) studies (Talbot et al. 2015).

A robust bioanalytical assay strategy enables suitable decision-making and interpretation of PK, PK/PD, efficacy and safety studies. An appropriate bioanalytical strategy could involve multiple bioanalytical tools and technologies, and a combination of qualitative and quantitative methods to understand the impact of biotransformation of mAb biotherapeutics on their pharmacological function. For instance, the anomalous PK profile warrants further investigation into the mAb biotherapeutic structure by orthogonal technologies/platforms. Liquid chromatography coupled mass spectrometry (LC-MS) platform is routinely used to troubleshoot for the anomalous immunoassay results. Similarly, loss in the pharmacological activity of mAb can be investigated by post-translational modifications (PTMs) analysis.

The bioanalytical assay employed for regulated non-clinical and clinical study support should adhere to FDA/EMA mandated guidelines for bioanalytical validation and should include validation of reference standards and critical reagents, optimization of calibration standards and limit of quantitation, minimal required dilution determination, specificity, selectivity, dilutional linearity, prozone or hook effect, and analyte stability. Parallelism analysis helps determine the matrix interference during the in-study sample analysis and should be evaluated for assays for endogenous compounds (FDA 2018; M10 2019; Medicines Agency 2011). The changes to the assay parameters due to the lot-to-lot variability of drug product and/or the biological matrix being used for the assay validation should be supported by strong scientific logic to justify the change(s) in the proposed plan for the assay validation. A detailed report and appropriate documentation for the validated assay should be prepared under GLP guidelines. A bioanalytical report summarizing the study samples results is prepared at the end of study sample analysis.

The investigational new drug-enabling nonclinical studies and clinical trials are supported by validated assays. A partial assay validation is warranted when slight modifications to the assay are made. The changes in the manufacturing lot number of the mAb biotherapeutic drug, critical reagents, biological matrix as well as co-administration of other drugs would warrant partial validation of the assay to demonstrate assay robustness. The assay transfer to another site/laboratory requires cross validation of the assay. The assay validation for monospecific mAb biotherapeutic normally requires validating one PK mAb analyte, but as the complexity of mAb modality increases, e.g., bispecific, trispecific Ab, then assay may require quantitating multiple PK domain-specific mAb analytes.

2.3.2 *PK Quantitation by LC-MS/MS*

Hybrid LBA based LC-MS/MS platform is becoming increasingly prevalent in the field of quantitative bioanalysis of biotherapeutics. LBA-LC-MS/MS overcomes several LBA-related shortcomings, such as cross reactivity, interference, critical reagent availability/consistency, and time-consuming critical reagent generation activities and assay development timelines. In addition, LC-MS/MS enables detection of degradation products and posttranslational modifications and distinction between highly homologous analytes. LC-MS/MS multiplexing capability is another superior advantage for quantification of co-administered biotherapeutics, and it has shown ability to provide valuable information about structural integrity and detection of different biotherapeutic domains. Flexibility and wide applicability of different LC-MS strategies enable quantification and/or characterization of intact proteins, subunit fragments, and signature peptides. Recent MS instrumentation with significantly enhanced sensitivity overcomes the assay sensitivity issue related to the production of multiple m/z precursor ions in the source.

Different LC-MS approaches can be applied for quantification and/or characterization of intact mAb therapeutics (top-down approach), subunit fragments (middle-up approach), or signature (surrogate) peptides (bottom-up approach). Due to the structural complexity and high molecular weight of mAbs, as well as the complexity of biomatrices, bottom-up approach with immunoaffinity capture is the most reliable workflow for accurate and sensitive quantification, and it is the most practical approach for assessment of mAb therapeutics efficacy and safety during nonclinical and clinical studies. Using proteolytic peptides with unique amino acid sequences, proper peptide length, and appropriate physicochemical properties as surrogates to measure intact mAb concentrations has been found to be a very successful strategy and will be the focus of this chapter.

Despite the structural complexity of mAbs compared to small molecules, similarities within this class of proteins have allowed for accelerated assay development time. All therapeutics mAbs are IgGs (e.g., IgG 1 or IgG 4) containing heavy and light chains, with the heavy chain determining the subclass of each antibody and the constant regions being conserved within each isotype. This unique structure across various types of IgG mAb therapeutics enabled scientists to develop different LC-MS/MS workflows with different levels of selectivity and sensitivity (Fig. 2.2). Common immunoaffinity capture and common protein digestion workflows enhance the wide applicability and reduce development time for LBA-LC-MS/MS mAbs bioanalysis.

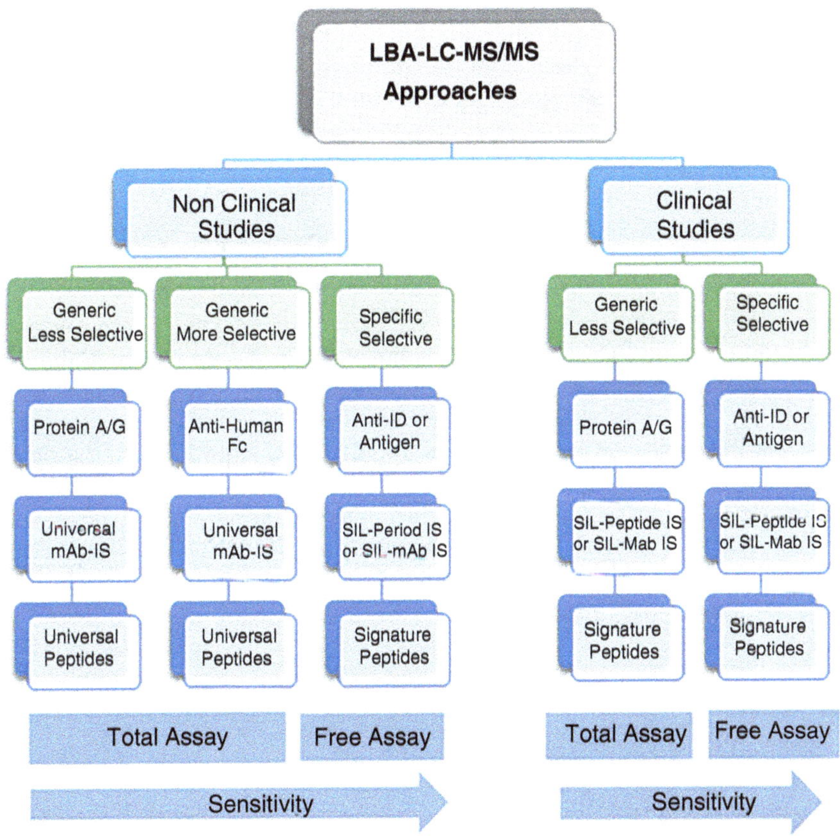

Fig. 2.2 Overall LBA-LC-MS/MS approaches for mAb quantification

2.3.2.1 Sample Preparation

Protein Precipitation (Pellet Precipitation)

Pellet precipitation followed by protein digestion is a simple, fast, and high-throughput workflow. Organic solvent (e.g., acetonitrile or methanol), perchloric acid, or salts are added to the sample to precipitate mAbs out of solution. Low-molecular-weight matrix components such as salts and phospholipids can be removed in the supernatant; however, other endogenous proteins such as albumin will be co-precipitated. Further treatment may be applied to selectively remove albumin in supernatant (e.g., isopropyl alcohol containing 1%trichloroacetic acid). The addition of organic solvents not only precipitates the protein but also helps in protein denaturation, allowing for direct pellet digestion without further denaturation and reduction. Surfactant-aided precipitation (i.e., adding surfactants into samples before precipitation) can be used for protein analysis in biometrics not only to enhance protein denaturing, reduction, and alkylation but also to improve peptide recovery and phospholipid removal. Protein precipitation followed by pellet digestion provides high digestion efficiency, and it is appropriate for total mAb assay. However, the main drawbacks of this sample preparation technique are the endogenous proteins co-precipitation and co-digestions, which may lead to interferences in some assays, especially those needing higher sensitivity (An et al. 2015; Becher et al. 2017; De Jong et al. 2020).

2.3.2.2 Protein Immunocapture

Immunocapture is the assay of choice for purification and enrichment of target mAbs from biofluids. This technique adds another layer of selectivity to the LC-MS assay. Different immunocapture strategies can be applied based on various factors such as (1) type of the assay (clinical vs nonclinical), (2) aims of the assay (free (unbound mAb) vs. total mAb measurement), (3) type of target biotherapeutic (e.g., capture target based on IgG structure such as with mAbs, or capture target based on the human Fc region such as with mAbs and fusion proteins), and (4) the required level of selectivity and sensitivity. Immunoaffinity purification has also been adapted to magnetic beads technology, 96-well format, and automated processing to improve the assay throughput.

Generic Immunocapture

Nonclinical Studies

Generic immunoaffinity purification is a commonly used assay for therapeutic mAbs sample cleanup. Purification of mAb candidates in different animal biomatrices can be done by using generic, readily available reagents such as protein A and protein G, and targeting constant (non-unique) regions such as the Fc region. One capture reagent can be used individually or simultaneously to enrich different mAbs. Generic immunocapture approaches are simple and almost development-free protocols with wide applicability. The same assay conditions may be readily adapted for various mAbs therapeutics. However, these approaches are considered less selective and result in less sensitive assays. Protein A and G or a combination of protein A/G has varying binding affinities to different IgG types and subtypes from different species, and therefore endogenous IgGs present in animal samples will be captured along with the target biotherapeutic. Binding capacity should be optimized carefully based on sample volume and animal species. If less sensitive assay was needed, Protein A and Protein G may be used, as an LLOQ of ≥1.0 μg/mL can be readily achieved for various mAbs therapeutics (Kaur et al. 2016; Lee 2016).

For more sensitive assay, a more selective generic immunocapture approach such as antibodies against human Fc region (Anti-Human Fc) Ab) can be used. This approach is also based on commercially available capture reagents but with a more specific immunocapture mechanism. Anti-human Fc Ab specifically targets the Fc region of human IgG with no cross affinity to other species. Biotinylated anti-human Fc Ab with streptavidin beads can be used for more selective purification/enrichment of mAbs and fusion proteins in animal biofluids and tissues. Lower LLOQs (e.g., 100 ng/mL) (Kaur et al. 2016) can be achieved with this approach. However, anti-human Fc Ab can be used only for nonhuman samples. Generic immunocapture reagents such as protein A/G and anti-human Fc Ab bind to both free and bound mAbs and are applicable for total mAb assay.

Clinical Studies

Generic captures using Protein A and Protein G can be also used for human studies. Low selectivity/sensitivity, elevated background, and binding capacity are the main limitations as described previously for nonclinical studies.

Selective Immunocapture

More selective immunocapture purification using Anti-ID antibodies or Anti-ID antibody fragments can be used for both clinical and nonclinical assays. Anti-IDs specifically bind to the idiotypic domain (variable region) of mAb therapeutics, mostly the CDRs removing almost all other endogenous IgGs, proteins, and other matrix components during washing. This approach provides cleaner samples and improves the assay sensitivity significantly (i.e., 100-fold or more) in comparison to generic immunocapture approach (Fernández Ocaña et al. 2012).

Antigen or ligand immunocapture is another selective purification approach, which is based mainly on using mAb targets as immunocapture reagents (e.g., vascular endothelial growth factor type A (VEGF-A) or tumor necrosis factor-α (TNF-α)). Ligands selectively bind to mAb targets in biofluids. One ligand may be used to extract different therapeutics; for example, TNF-α can be used to extract both infliximab and adalimumab (Jourdil et al. 2018). Ligand-immunocapture approaches are more selective than generic immunocapture but less selective than the anti-ID approach, as ligands may bind multiple proteins in biomatrices.

Anti-IDs and ligands have been used highly effectively to extract different mAb therapeutics from biomatrices (DelGuidice et al. 2021; Dubois et al. 2008; Fernández Ocaña et al. 2012; Xu et al. 2014). However, sample purification with selective reagents can be costly and is not applicable in each case. Additionally, selective reagents bind only free mAbs or partially free mAbs (at least one Fab arm should be available for binding). mAbs bound to soluble targets or anti-drug antibodies may be lost during sample purification, if additional dissociation steps are not employed. For this reason, two different assays using selective (e.g., anti-ID) and generic (e.g., protein G) immunocapture may be needed for free and total mAb PK studies, respectively (Fernández Ocaña et al. 2012).

Anti-IDs, ligands, and anti-human Fc Ab are usually labeled (e.g., with biotin) and selectively immobilized on magnetic beads (e.g., streptavidin, Strep-Tactin coated magnetic beads). The capture order (i.e., incubation of mAb with capture reagent followed by immobilization onto beads or immobilization of capture reagent on beads followed by mAb capture incubation) is usually evaluated and compared during assay development. Target mAb to capture reagent ratios (e.g., 1:8, 1:10 or 1:12) should also be compared and optimized.

2.3.2.3 Protein Digestion

The ability to have a common workflow is one of the main advantages of using LBA-LC-MS/MS for mAb bioanalysis; with a typical enzymatic digestion usually being applied after the immunocapture step but before LC-MS/MS analysis. Efficient and optimized enzymatic digestion is mandatory to reproducibly and reliably generate surrogate peptides. The enzymatic digestion protocol includes a few main steps that need to be optimized to ensure efficient and reproducible protein digestion. Denaturation using heat and/or addition of surfactant (such as RapiGest™ or Protease MAX™) is firstly applied to unfold the tertiary structure of mAb followed by or done concurrently with a reduction step using a reducing agent such as Tris(2-carboxyethyl)phosphine hydrochloride (TCEP) or dithiothreitol (DTT), to reduce the disulfide bonds between the heavy and light chains and make the molecule fully accessible to the digestion enzyme. The next step will be an alkylation step using alkylating agents, such as iodoacetic acid (IAA) or iodoacetamide (IAM), to block the reactive free thiols group and prevent the disulfide bonds reformation. After alkylation, the mAb protein can then be digested. The selection of digestion enzyme is based on the amino acid sequence of the target surrogate peptides. Trypsin is a commonly used enzyme that cleaves peptide chains at the carboxyl side of amino acids lysine and arginine. It is frequently used because it provides reproducible digestion efficiency and generates tryptic peptides with proper length (i.e., suitable for MS detection). Other enzymes such as Asp-N®, Glu-C®, Lys-C®, protease K, papain, and pepsin can be used alternatively or sequentially with tryptic digestion to generate unique peptide sequences and to improve assay selectivity and sensitivity. The enzyme-to-protein ratio is critical to ensure highly reproducible protein digestion, with a 1:20 or higher enzyme/protein (w/w) ratio recommended for adequate surrogate peptide yields. Digestion temperature and digestion buffer pH are also key factors for digestion efficiency (e.g., 37 °C and digestion buffers (pH ~ 8) are commonly used for tryptic digestion). Incubation time is based on the position of target surrogate peptides and accessibility to digestion enzymes. Longer incubation times, such as overnight digestion, which were previously commonly used may potentially induce amino acid changes such as asparagine deamidation, N-terminal glutamine cyclization, and methionine oxidation. Shorter digestion durations (~1–3 h) are currently more common and have been seen as more effective, using fast and ultrafast digestion. Fab selective proteolysis for mAbs (nanosurface and molecular-orientation limited (nSMOL® proteolysis) using nanoparticle-immobilized trypsin is a recent approach limiting protease access to the FC region and minimizing sample complexity (Iwamoto et al. 2016; Wang et al. 2016; Yuan et al. 2019).

2.3.2.4 Surrogate Peptide Selection

While typical protein digestion is considered a straightforward protocol with some required optimizations, selection of surrogate peptide is a complicated but very critical step for accurate quantitation of mAb therapeutics. Surrogate peptide selection is based on two main factors: whether the study is clinical or nonclinical and the type of mAb therapeutic (e.g., chimeric, humanized, or fully human mAb). In silico tools, such as Skyline, are usually applied to identify potential peptide candidates, which are usually then filtered (to determine uniqueness) against the corresponding study species proteome (using different software tools and databases such as BLAST® and UniProtKB/Swiss-Prot, respectively), followed by LC-MS/MS experimental testing to select the optimal surrogate peptides and to ensure absence of interference from the study biomatrix. The initial in silico selection may exclude peptides with improper length and peptides containing amino acid sequences that are susceptible to in vitro or in vivo changes such as oxidation, deamidation, isomerization, and cyclization. The experimental testing usually excludes candidates with poor chromatographic retention/peak shape, inadequate MS ionization, or selectivity/stability issues.

Once surrogate peptides with the optimal LC-MS/MS performance (i.e., peak shape, chromatographic retention, MS ionization, sensitivity, specificity, and stability) are finally identified, additional surrogate peptides from different antibody domains can be identified for quantitation and confirmatory purposes. Digestion efficiency, chromatographic conditions, and MS parameters are evaluated and optimized to improve assay sensitivity and ruggedness.

Universal Surrogate Peptides

A generic approach using universal surrogate peptides from the constant region of human IgGs is widely used for quantitative bioanalysis of human/humanized mAb therapeutics in animal matrices. The amino acid sequences of the human IgG constant region are unique and different from non-human IgGs. This approach showed wide applicability in early preclinical studies to evaluate mAbs in different animal species and nonclinical evaluation of multiple mAb candidates.

Universal surrogate peptides for different human IgG classes have been previously identified (Furlong et al. 2012) and successfully applied for quantification of different mAb therapeutics in different animal matrices. Universal surrogate peptides from heavy chain and light chain constant regions can be used to ensure mAb structure integrity and to distinguish between the structurally intact molecule and degraded forms (Furlong et al. 2013). Minimum or no extra development work may be required for individual mAb therapeutics, since the same chromatographic conditions and MS parameters can be applied as the final target (i.e., surrogate peptide) are the same. Universal surrogate peptide approach has been successfully applied for different nonhuman PK studies, including a PK profile obtained from the universal peptide approach in monkey plasma samples which was in excellent agreement with the data obtained from the unique signature peptide approach and an ELISA (Kaur et al. 2016). Another monkey PK study showed that the data obtained from both light-chain and heavy-chain universal peptides were also with excellent agreement, which not only supported the applicability of the approach for quantitation but also confirmed the in vivo structural integrity of target mAb therapeutic (Furlong et al. 2013). LBA-LC-MS/MS assays for seven different mAb therapeutics in monkey, rat, and mouse biofluids were developed and validated using universal surrogate peptides for both quantitative and confirmatory purposes (Kaur et al. 2016).

This universal approach is only applicable for nonclinical samples and not suitable for non-humanized mAbs or assay with simultaneously co-dosed mAbs from same IgG subclass, or different subclasses contain same universal surrogate peptides (e.g., VVSVLTVLHQDWLNGK peptide is a conserved peptide in the Fc region of human IgG1 and IgG4))(Furlong et al. 2013).

Signature Surrogate Peptides

Using surrogate peptides with unique amino acid sequences from the CDRs is a highly specific quantitation approach for mAb therapeutics, since CDRs are variable regions unique to each individual mAb. This approach can be applied for both clinical and nonclinical studies and is applicable for simultaneous determination of co-administered mAb therapeutics. This approach improves assay selectivity and sensitivity significantly due to the uniqueness of surrogate peptides. However, since a new assay should be developed for each individual mAb therapeutic, this can increase development time. Due to the smaller size of the variable region in comparison to the constant region, finding signature peptides with optimal LC-MS/MS performance may be a challenge in some cases. Examples of analytical challenges related to limited signature peptide choices include (1) more than one enzyme needed to improve digestion efficiency (e.g., tryptic hydrolysis hindrance due to presence of proline), (2) stabilization of unstable peptide candidates (e.g., optimization of temperature and pH during sample digestion to minimize asparagine deamidation and using high-resolution MS to distinguish between amidated and non-amidated forms), and (3) use of peptide candidates with methionine that is prone to oxidation, which will require monitoring the presence of both oxidized and non-oxidized forms, potentially adding stabilization steps, such as degassing reagents, working under reduced light/temperature, and adding antioxidants (i.e., methionine or sodium thiosulfate) to samples, to limit the effect on assay sensitivity.

2.3.2.5 Internal Standard

Using an internal standard (IS) is essential for accurate and reliable LC-MS/MS quantitation. Stable isotopically labeled internal standards (SIL-IS) are commonly used for quantitative LC-MS/MS. SIL-IS are very similar to target analytes in terms of physicochemical properties, chromatographic retention, and MS ionization, with enough mass difference. SIL-IS usually tracks variabilities related to sample processing (e.g., recovery) and MS detection (e.g., suppression or enhancement due to endogenous matrix components). While using SIL-IS is mandatory in LBA-LC-MS/MS, there are different internal standard types available (Fig. 2.3) that can be added at different steps of sample processing as described below.

Stable Isotope Labeled Peptide Internal Standard

SIL-peptide ISs are the most commonly used SIL-ISs in mAb bioanalysis. SIL-peptide ISs contain the same amino acid sequence as the surrogate peptides with one or two amino acids labeled with ^{13}C and ^{15}N, providing enough mass shift to distinguish surrogate peptides from the corresponding SIL-peptide ISs (Kaur et al. 2016). SIL-peptide ISs are added at the end of the extraction process before LC-MS/MS analysis and track any variabilities related to chromatography and MS ionization, especially matrix effects (i.e., ion suppression and enhancement) related to co-eluted matrix components or other peptides formed as a result of the protein digestion. SIL-peptides can be synthesized with high purity in a short time. However, variabilities related to immunocapture efficiency or mAb digestion cannot be compensated by SIL-peptide IS, so reproducibility of these steps should be evaluated during assay development to ensure assay ruggedness. Addition of SIL-peptide IS before digestion has been suggested; however, IS degradation or instability may occur, so caution should be employed (Arsene et al. 2008; Shuford et al. 2012).

Extended SIL peptides (SIL peptides with extra amino acids on one or both sides) (Barnidge et al. 2004; Faria et al. 2015) can be added before digestion to track

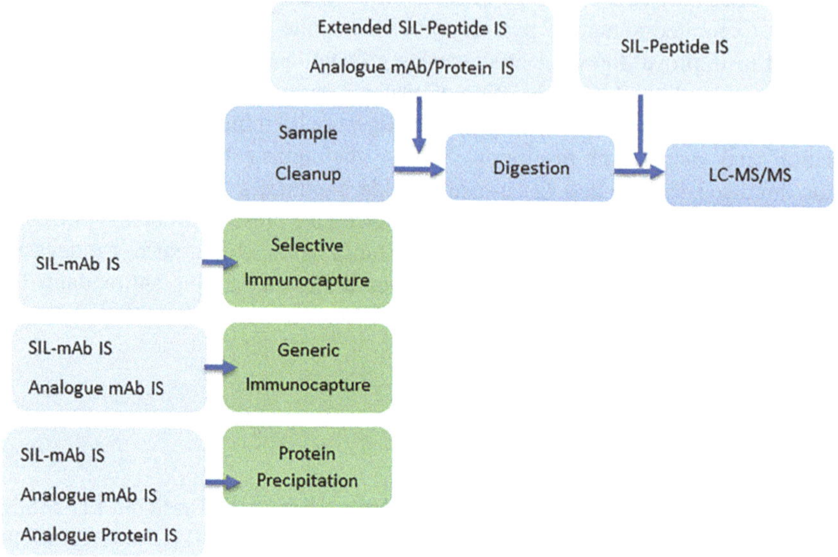

Fig. 2.3 Different types of internal standards are used for different workflows and can be added at different sample processing steps

digestion variability, as the extra amino acids will be cleaved during digestion to form SIL peptide. This may not be as effective at tracking as initially perceived, however, because the digestion kinetics may be significantly different between intact mAb and extended SIL-peptide IS due to differences in accessibility to the digestion enzyme and time required for digestion. SIL-peptide IS can be used for both nonclinical and clinical studies, and they are very convenient for simultaneous assays of multiple mAb candidates, or multiplex assays of surrogate peptides from different domains (e.g., heavy chain and light chain). Similar assay performance has been reported using SIL-mAb IS and SIL-peptide IS (Sucharski et al. 2018). SIL-peptide ISs are not only used for quantitative surrogate peptides but also can be used for confirmatory peptides. For example, SIL-peptide ISs have been used for quantitation of bevacizumab and its related fab fragment therapeutic ranibizumab, while a confirmatory SIL-peptide IS corresponding to the confirmatory ranibizumab surrogate peptide was simultaneously used to distinguish between both biotherapeutics in human plasma (DelGuidice et al. 2021).

Stable Isotope Labeled Monoclonal Antibody Internal Standard

A SIL-mAb IS is the ideal IS for LBA-LC-MS/MS, as it can be added to the sample at the beginning of sample processing and corrects all variabilities related to immunocapture efficiency, digestion efficiency, chromatography, and MS detection. ^{13}C and ^{15}N isotopes can be incorporated into essential amino acids (such as lysine, arginine, leucine, and valine). SIL-mAb ISs have been used for LBA-LC-MS/MS analysis of several mAb therapeutics (El Amrani et al. 2016). However, a new SIL-mAb IS should be synthesized for each assay, which may be not available in each case, due to high cost and time-consuming synthesis.

Universal SIL-mAb ISs, such as SiluMab™, are mAb-ISs containing human IgG heavy-chain and light-chain constant regions with stable isotope-labeled amino acids such as ([^{13}C6,^{15}N]-l-leucine and [^{13}C5,^{15}N]-l-valine). They are commercially available and are widely used in nonclinical studies (e.g., SIL human IgG1 and IgG4). Stable isotope-labeled universal peptides are generated from trypsin digestion and can be used to track the corresponding universal surrogate peptides. Universal SIL-mAb ISs have been reported for quantification of several mAb therapeutics in nonclinical studies (Furlong et al. 2012; Kaur et al. 2016).

Analogue mAb-ISs or analogue protein-ISs may also be used, which can be added during sample processing (e.g., pellet precipitation) or before digestion to track variabilities related to the digestion process. Differences in digestion efficiency between target mAb and Analogue mAb-IS or analogue protein may impact the assay ruggedness. Selection of unique peptide IS generated after digestion is critical, with selection of amino sequences and peptide locations in both target mAb and analogue being more critical than peptide retention times (Halquist and Karnes 2011; Ismaiel et al. 2017; Osaki et al. 2017).

2.4 Immunogenicity

The mAb biotherapeutics are prone to elicit unwanted immune responses against themselves. The unwanted immune response can alter or reduce the efficacy, PK/PD and may be associated with adverse effects. This section focuses on ADA based immune response against mAb biotherapeutics. Detection of ADA is typically carried out using qualitative and/or quasi-quantitative assays because of the unavailability of standardized, species-specific (especially human) polyclonal ADA reference materials to use as assay calibrators (Findlay et al. 2000).

2.4.1 Immunogenicity Assessment by Immunoassays

A multi-tiered approach is typically used to evaluate immunogenicity against mAb biotherapeutic as illustrated in Fig. 2.4. The standard tiered strategy involves Screening, Confirmatory and Titer assays to detect, confirm and quasi-quantitate ADA against mAb biotherapeutic. Depending on the immunogenicity risk assessment, further characterization (isotyping, neutralization activity, epitope mapping etc.) may be needed. Typically, a homogeneous bridge assay format employing labeled mAb biotherapeutic as capture (e.g., biotin-labeled mAb

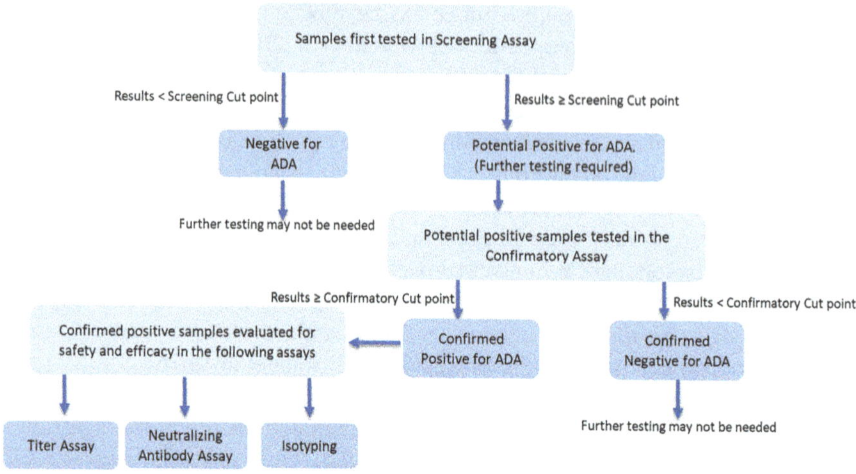

Fig. 2.4 Multi-tiered approach for immunogenicity testing against mAb biotherapeutics

biotherapeutic) and detection (e.g., ruthenium labeled mAb biotherapeutic) reagents is used for immunogenicity assays.

2.4.1.1 Screening Assay

The samples are first tested in the screening assay to determine if they are potentially positive or negative for ADA against mAb biotherapeutic. The screening assay format is selected to detect both low- and high-affinity ADA as well as all the subclasses of ADA.

Screening Cut Point

The screening cut point (SCP) is the assay threshold response at or above which a sample is considered to be potentially positive for ADA. An assay response below the screening cut point is considered negative for ADA. The screening cut point is statistically determined taking into account a defined false positive rate of approximately 5% to maximize detection of true positives in the assay.

The SCP is estimated by testing appropriate number of treatment-naïve samples (≥10 individual subjects during assay development and ≥50 individual subjects during assay validation) from the intended subject population, wherever possible or using the healthy population. Each sample is tested by at least two analysts on at least three different days for a total of at least six individual measurements (FDA 2019). During later phases of clinical studies based on the availability, cut points should be determined again using diseased state samples from the actual patient population. For oncology indications, if a drug is being used to target multiple cancers, then at least 20 subjects from each of these disease sub-populations can be included in cut point evaluation (Devanarayan et al. 2017).

The data obtained from the cut point runs are evaluated for normality using a commonly employed, Shapiro Wilk test. If the data is found to be non-normal, it is first log-transformed. The normal distributed or log-transformed data is then assessed for biological (a sample with signal that is consistently very high or low when compared to the other samples being evaluated for cut point) and/or analytical (one sample signal that is abnormally high or low when compared to all the sample signal results obtained for that sample) outliers using commonly employed Whisker box plot analysis or ANOVA.

After removal of outliers, the data is re-assessed for normality. A) If the data is normally distributed, then the mean and standard deviation (SD) of the samples is used to calculate the assay cut point as shown below.

Mean + 1.645*SD (Shankar et al. 2008).

If log transformation was used to normalize the data, then the log values of mean and SD are used to calculate the assay cut point as shown below.

Antilog([logMean] + 1.645* [log SD]).

This approach is called the parametric approach.

A variation of the parametric approach involves the use of median for assay cut point estimation as shown below.

Median + 1.645 × (1.483 × MAD), where MAD is median absolute deviation (Shankar et al. 2008).

The MAD is calculated as follows:

1. Calculate median of all data.
2. Record absolute values obtained by subtracting all cut point sample results from this median.
3. The median of absolute values obtained in "2" above is the MAD.

Another robust alternative for assay cut point estimation is the Tukey's biweight procedure (Mosteller 1977).

B) If after the outlier evaluation, the data is found to be distributed non-normally then non-parametric approach can be used. In this approach, the 95% percentile of all the sample is used to estimate the assay cut point.

2.4.1.2 Confirmatory/Specificity Assay

In the confirmatory or specificity assay, samples that screened as potential positives are further tested to confirm their specificity against mAb biotherapeutic. Typically confirmatory assays employ competition inhibition assay strategy, wherein samples are evaluated in the presence and absence of excess unlabeled mAb biotherapeutic. The excess amount of unlabeled mAb biotherapeutic is carefully and experimentally chosen in a manner that it does not result in atypical results for high- and low-positive ADA samples.

In the competitive inhibition assay, excess unlabeled mAb biotherapeutic competes with the labeled mAb biotherapeutic (capture/detection) reagents for binding to the ADA in the sample. As a result, in the presence of excess unlabeled mAb biotherapeutic, the ADA signal specific to its binding to mAb biotherapeutic is inhibited. If the percent inhibition is at or above the established assay threshold (confirmatory assay cut point) then the sample is confirmed positive for ADA. Otherwise, it is confirmed negative for ADA.

Confirmatory/Specificity Cut Point

The confirmatory assay cut point determination is carried out in a manner similar to the screening cut point determination. The samples from healthy or drug-naïve subjects used for screening cut point determination are spiked with an excess known amount of mAb biotherapeutic, and the %inhibition of assay signal is evaluated using the following formula:

$$\left(1-\left(\text{Drug Inhibited sample} / \text{Uninhibited sample}\right)\right)*100$$

The %inhibition data is analyzed for normality and outliers (biological and analytical) similar to what is done for screening cut point determination (see previous section). Unlike the screening cut point, the confirmatory cut point is calculated taking into consideration 1% false positive rate to reduce false positives generated in the screening assay.

After outlier removal, the % inhibition data is evaluated for normality distribution assessment. A) If the data is found to be distributed normally then parametric approach is used as shown below.

- (1 – (Mean – [2.33*SD])) *100.
  If the log-transformation on the data was performed then the log values of Mean and SD are used to calculate confirmatory cut point as shown below.
- (1 – Antilog(log Mean – [2.33 * (logSD)])) *100.
  B) If the data is found to be distributed non-normally then non-parametric approach is used, wherein the value of the 99th percentile of all the samples is used to calculate the confirmatory cut point.

2.4.1.3 Titer Assay

A quasi-quantitative assay is performed to determine the titer for ADA. In the titer assay, confirmed positive samples are serially diluted and tested in the screening assay format. The reciprocal of the highest dilution of the sample (including MRD) that yields a positive result is reported as the titer value for that sample.

The titer cut point is generally calculated using data from the screening assay cut point determination.

2.4.1.4 Isotyping Assays

Regulatory agencies recommend that if there is a high risk of anaphylaxis or other unwanted immunogenic response then Ig subclass determination and/or T-cell reactivity determination may be carried out. The development of Ig subclass specific ADA assay requires the generation of Ig-specific detection antibodies and/or positive controls.

2.4.1.5 Neutralizing Antibody Assay

Neutralizing antibody (Nab) assays help detect ADA that neutralize the pharmacological effect of the mAb biotherapeutic. The selection of the appropriate assay format for Nab assay depends on the mAb biotherapeutic's mechanism of action, its immunogenicity risk, and the assay performance characteristics (Wu et al. 2016).

Two types of assay formats are used to detect neutralizing antibodies: cell based and non-cell based assays. For each of these two categories, both indirect and direct assay format can be used to detect NAbs.

Direct Assay

For cell-based assays, samples are incubated with cells that express the receptor to which the drug will bind. So, in the presence of a higher amount of NAb, neutralizing antibodies will bind to regions of the mAb relevant for therapeutic action, resulting in neutralization via steric hindrance and a subsequently lower cellular response to the drug and vice versa. For non-cell-based assays, the mAb is preincubated with the NAb or samples suspected of having NAbs. The NAb will bind to the mAb preventing it from binding to anything else that might be used to detect it (e.g., labeled drug target). This means that the raw signal produced will be inversely proportional to the NAb levels (Wu et al. 2016).

Indirect Assay

For cell-based assays, samples are incubated with a ligand and cells that have receptors specific to that ligand as well as the drug. So in the presence of a high amount of NAb, the drug will bind to it, while the ligand will bind to the cell and produce a higher cellular response, but in the absence of NAbs, the drug will bind to the ligand and nothing will bind to the cell leading to zero or low cellular response. For non-cell based, a ligand that binds with both a target receptor and the drug is chosen. The ligand is preincubated with samples. If the samples have a high incidence of NAbs, then they will bind to the drug and allow the ligand to bind to the target receptor and produce a higher raw signal. But, if there is a lower incidence of NAb, then the drug will bind to the ligand which will in turn not bind to the receptor and produce a low signal (Wu et al. 2016).

The FDA recommends that, for cut point determination, a minimum of 30 samples be tested on three different days by at least two analysts (FDA 2019).

Data from these 30 samples is analyzed for normality and outliers using the strategies discussed in the screening and confirmatory assay portion of this chapter.

2.4.2 Analytical Considerations for ADA Assays

2.4.2.1 Sensitivity and Positive Control

Assay sensitivity is the lowest concentration at which the antibody preparation consistently produces either a positive result or a readout equal to the cut point determined for that assay (FDA 2019). Evaluation of assay sensitivity is carried out using responses obtained from positive control antibodies (PCs). They can either be polyclonal or monoclonal antibodies generated using multiple techniques like animal immunization, phage display etc. The FDA recommends that these PCs be affinity purified using the therapeutic protein product (FDA 2019). This will aid in making the PC more specific, especially if it's a polyclonal Ab, to the ADA and will lead to an increase in sensitivity. The animal species that the PCs were generated in is another aspect of PC development that can affect ADA assays and sensitivity. The reason is that the detection/capture/bridging antibodies used in the ADA should be able to bind to not just these PCs but also human ADAs.

Once a PC is selected, sensitivity assessment is carried out by testing multiple serial dilutions of the PC in naïve matrix obtained either from an individual or a matrix pool. A minimum of five serial dilutions, which are limited to two- or three-fold serial dilutions, of the PC are tested. The results are plotted on a curve, and the linear portion of the curve that corresponds to the assay cut point is determined to be the sensitivity of the assay. Reporting of assay sensitivity is therefore done as a mass of antibody detected per milliliter of undiluted matrix. This means that assay sensitivity measurements must include the assay's minimum required dilution (MRD) (FDA 2019). So, if an assay with an MRD of 10 has an assay sensitivity of 10 ng/mL, then the assay sensitivity is reported as 100 ng/mL.

The FDA recommends that screening and confirmatory IgG and IgM ADA assays achieve a sensitivity of at least 100 ng/mL (FDA 2019). This is because concentrations as low as 100 ng/mL have been known to be associated with clinical events (SA 2010; Zhou et al. 2012). For ADA assays that detect IgE, ADA should have sensitivity in the high picograms per milliliter (pg/mL) to low ng/mL range (FDA 2019). It should also be understood that neutralization assays may not achieve these levels of sensitivity. In such circumstances, regulatory agencies may ask to see proof that attempts were made to develop assays that were more sensitive.

2.4.2.2 Drug Tolerance

Drug tolerance is the ability of an assay to achieve similar levels of sensitivity both in the presence and absence of a certain quantity of drug. Sensitivity measurements are generally complicated by the fact that the presence of the therapeutic in the blood, and therefore samples, will lead to inaccurate detections of ADA. This is because the competition between the drug and the capture/detection reagents from the assay system for product-specific antibodies or ADA will lead to false-negative results (Shankar et al. 2008). This problem is further intensified when the therapeutic is a monoclonal antibody. mAbs continue to exist in the body for a long time and therefore have long half-lives. So, depending on when the blood samples are collected or drawn from patients, these existing mAbs can bind to ADAs and cause reporting of false negatives. These factors in turn affect the efficacy and safety profile of the drug.

Many strategies are employed to achieve the expected drug tolerance. One possible way is to induce the ADA-drug complex to break by using an acid treatment. This assay is called acid dissociation. Other assays employed to enhance drug tolerance are solid-phase extraction with acid dissociation (SPEAD), affinity capture elution (ACE), bead extraction and acid dissociation (BEAD), precipitation and acid dissociation (PandA), etc. Drug tolerance of an assay also depends on the assay platform and the quality of the reagents being used.

Achieving the expected drug tolerance in an ADA assay is therefore a delicate dance to find a balance between the ideal strategy/assay, assay platform, and reagents.

2.4.2.3 Preexisting Antibodies

For many drug therapies, even before its administration, the human body may already have antibodies against the drug. These preexisting antibodies will affect the PK/PD of the mAb. They will also affect the sensitivity and understanding of the results obtained from ADA assays. For a mAb with a preexisting antibody, the screen, confirm, and titer value might be very high because it will be indicative of the ADA response to the mAb therapy in addition to the preexisting antibody. So different, non-traditional approaches that are cognizant of these factors are used to make the immunogenicity data more relevant. The most common approach is to assess ADA responses in an individual subject's results before and after treatment (FDA 2019). So, if for example, a subject's titer result is two dilution steps greater than the pre-treatment titer value, then it can be reported as treatment-boosted ADA response.

2.4.2.4 Target Interference and Rheumatoid Factor

Drug targets and rheumatoid factor (RF) are frequently the cause of interference in ADA detection. When a sample is tested in an ADA assay where the drug or labeled drug (biotin drug, sulfotag drug, etc.) is used to detect, capture, or form a bridge (bridging assay format), then the drug binds to the drug target instead of the ADA creating a false-positive signal. Rheumatoid factors (RFs) are endogenous human antibodies that bind to human gamma globulins (Tatarewicz et al. 2010). So RFs will bind to the mAb drug and cause false-positive signals. ADA assays for mAbs are therefore particularly susceptible to interference by RFs.

Strategies like sample treatment with Melon Gel Resin purification, Protein G purification, etc. are employed to reduce interference by both targets and RFs. Assay formats and reagents also play a vital role in reducing false-positive measurements due to target inference and RFs.

2.4.3 *Immunogenicity Assessment by LC-MS/MS*

LC-MS/MS has emerged recently as a promising technology to supplement existing assays for immunogenicity testing. Due to specificity, selectivity, and multiplexing capability of LBA-LC-MS/MS and the well-established immunocapture and digestion workflows for biotherapeutics, different LC-MS strategies can also be applied for ADA testing. The main challenges associated with traditional ADA detection are drug tolerance, soluble target interference, matrix effects, reagent availability, and the time-consuming process to develop different assays for ADA screening, confirmatory, isotyping, and magnitude (titer). LC-MS technology showed promising capability to overcome some of these challenges such as high drug tolerance (i.e., ADA responses to therapeutic proteins can be measured in the presence of high concentrations of biotherapeutics); single assay can be used for ADA isotyping and semi-quantitation, and interference from target mAb therapeutic or endogenous IgG can be also simultaneously monitored (Chen et al. 2016; Jiang et al. 2014; Neubert et al. 2008).

However, more challenges and limitations are still encountered based on the assay format.

2.4.3.1 LC-MS Strategies for ADA Quantification

LC-MS strategies for ADA quantification are based on many factors such as (1) the type of ADA (e.g., IgG, IgM, or IgA), (2) the nature of the target biotherapeutic (e.g., IgG based, human Fc based, non-human Fc protein), and (3) the type of the assay (e.g., clinical or nonclinical). Due to consideration of assay ruggedness and sensitivity, the commonly used LC-MS approach for biotherapeutics is typical immunocapture followed by enzyme digestion and quantification of signature peptides derived from the target biotherapeutic. Same workflow is also common for ADA testing with some modifications (Fig. 2.5). LC-MS strategies for ADA may be generic workflows with some applicability or case-specific strategies for developing "fit-for-purpose" bioanalytical approaches with limited applicability.

2.4.3.2 LC-MS Strategies with Limited Applicability

Accurate quantification of free ADA is subject to interference from high concentrations of the biotherapeutic (i.e., assay tolerance to the presence of excessive biotherapeutics). High concentrations of the biotherapeutic may be added to the ADA

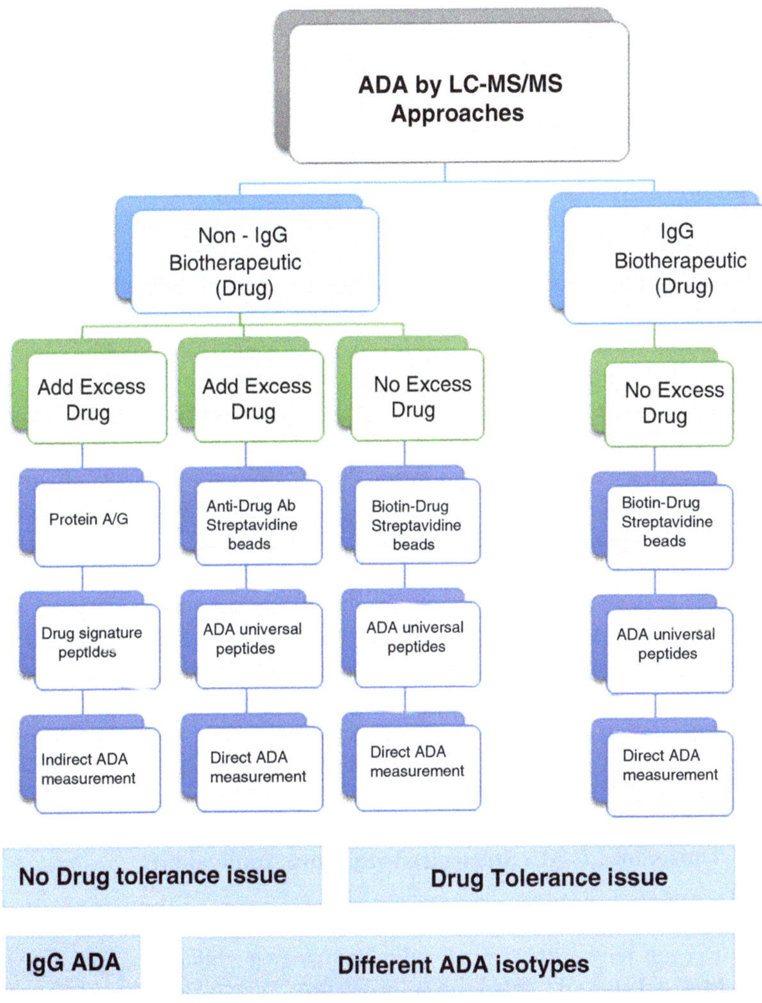

Fig. 2.5 LBA-LC-MS/MS approaches to ADA testing

samples to saturate ADA binding sites (i.e., bind all free ADA molecules) and to form ADA-biotherapeutic complexes. An immunoprecipitation step is applied to isolate ADA-biotherapeutic complexes followed by wash steps, elution, and digestion. The quantified signature peptides of the target biotherapeutic can be used as surrogates for the presence of ADA (indirect measurement). Generic capture reagent such as Protein A or G can be used for immunoprecipitation; however, this limits the detection to IgG-based ADA, and other capture reagents should be included (e.g., anti-human IgM) to bind other immunoglobulin subclasses. This approach is applicable for both clinical and nonclinical studies; however, affinity to IgGs from different species should be considered and evaluated. This approach is limited to non-IgG biotherapeutics (i.e., not applicable for mAbs or fusion proteins) as extra drug should not bind to protein A/G and should be removed during the immunocapture step, and only ADA-biotherapeutic complexes will be isolated (Neubert et al. 2008). New assay should be developed for each individual biotherapeutic candidate, and in silico study should be applied to select signature peptides from biotherapeutic, which are usually filtered against the study species proteome (human or non-human) and the ADA sequences, as both ADA and biotherapeutic will be digested and a pool of protein digest (i.e., peptides) will be produced.

Bead extraction and acid dissociation (BEAD) strategy is commonly applied to isolate ADA from biomatrices using traditional techniques such as cell-based assays. LC-MS can be used as a supporting technique for simultaneous detection of the residual amounts of mAb therapeutic which is complexed with the neutralizing ADA and the residual endogenous IgGs as a marker for nonspecific binding of matrix components during BEAD extraction. Biotinylated target mAb can be used as a capture reagent to form (biotinylated-mAb)-ADA complex and then captured by the streptavidin beads, followed by elution with acidic buffer, neutralization, reduction, alkylation, and digestion. Signature peptides derived from ADA, mAb, and human IgG are used as surrogates for the corresponding analyte (Jiang et al. 2014). Availability of human positive control (i.e., human ADA against target mAb) is the main drawback of this strategy. ADA from animal species (e.g., mouse IgG against the mAb therapeutic) can be used as a positive control to substitute for human Nab; however, ADA recovery may be different (human ADA vs. animal ADA substitute), and surrogate peptides are also different. This approach demonstrates the multiplexing capability of LC-MS; it can be used to estimate the ADA recovery and the efficiency of removing endogenous matrix components and/or residual mAb therapeutic using the BEAD technique. However, it is considered a case-specific supporting strategy with limited applicability.

2.4.3.3 Universal LC-MS Strategies with More Applicability

Comparable to the universal surrogate peptides approach for biotherapeutic, unique peptides to each human immunoglobulin isotype (e.g., IgG, IgE, IgM, IgA) and subclass (e.g., IgG1, IgG2, IgG3, IgG4, IgA1, IgA2) can be identified and used for

ADA isotyping and quantification (Chen et al. 2016). The identified surrogate peptide(s) should be from the constant region (i.e., unique for each isotype/subclass and not present in other isotypes). High background due to residual endogenous immunoglobulin in the final extract is the main challenge for the universal ADA peptide assay. This approach can be applied for clinical and nonclinical studies, and species-specific ADA universal peptides should be identified (e.g., VVSVLTVTHQDWLNGK peptide is present in all cynomolgus IgG1-4 isotypes and can be used to detect the total IgG1-4 ADA; however, it is also present in some other cynomolgus endogenous proteins, and a selective immunocapture should be applied) (Roos et al. 2016).

The universal LCMS strategies for ADA can be applied to different biotherapeutic studies with minimum modifications. Unique ADA surrogate peptides are species specific; same protein digestion workflow and LCMS conditions can be used for different studies on the same species. Two different workflows can be applied for ADA isolation and quantification:

(i) Biotinylated protein (e.g., mAb) can be used as a capture Ab to enrich ADA and form a biotinylated mAb-ADA complex, after immobilization on streptavidin beads. The bound ADA molecules can be eluted and digested. The generated ADA universal peptides can be used as surrogates for quantification (direct measurement) (Chen et al. 2016). Drug tolerance should be evaluated during assay development; the acid dissociation step is usually applied to dissociate mAb-ADA complexes, followed by neutralization and addition of biotinylated mAb. The ADA to biotinylated mAb ratio should be optimized; the higher the biotinylated mAb concentration, the better the immunocapture efficiency, and both biotinylated mAb-ADA and mAb-ADA complexes may be formed. Although mAb-ADA complexes will not bind to streptavidin beads, high mAb concentrations in the ADA samples will impact the ADA recovery (i.e., drug tolerance).

(ii) Excess biotherapeutic can be added to the sample to saturate all ADA binding sites, and biotinylated anti-protein Ab (e.g., mouse anti-drug Ab) can be used to capture protein-ADA complexes. This approach overcomes drug tolerance issues; however, it is not applicable to IgG-based therapeutic (e.g., mAb) clinical studies as human universal ADA peptides are also produced from mAb tryptic digestion. mAb signature peptides (indirect ADA measurement) can be used in this case as described before.

Calibration curves are prepared using standard materials of different Ig isotypes/subclasses, spiked into blank samples after the immunocapture process, and then undergo the same digestion process. This approach is considered a semi-quantitation, as the ADA recovery (i.e., immunocapture step) cannot be monitored or compensated by calibration curve and affinity of different Ig isotypes cannot be evaluated. High background due to nonspecific binding of endogenous Ig proteins, non-Ig proteins, and residual biotherapeutic during immunocapture process is the main challenge for this approach.

2.5 Future Developments and Challenges

While biotherapeutics mAbs share the largest market share for large-molecule biotherapeutics, novel mAb-based biotherapeutic modalities are also gaining popularity. The development of these new mAb-based biotherapeutic modalities such as bi-, tri-, and multi-specific antibodies, antibody-drug conjugates (ADCs) would leverage clinical experiences gained with mAb biotherapeutics. Biotherapeutics based on antibody-based fragments (Fab, scFv) potentially offer the advantage of increased penetration (due to its smaller size), lack of Fc region (better stability), and reduced manufacturing costs.

A new approach of targeting tumors by the T-cell redirection mechanism has been exploited for drug development purposes. Chimeric antigen receptor (CAR)-T cell therapy is based on the genetically engineered fusion proteins (scFv from variable region of mAb) connected to the T-cell activating motif (Elverum and Whitman 2020; Sinclair et al. 2018).

ADCs, bi-specific antibodies, and cell therapy are discussed in more detail in the respective ADC, bi-specific and Fusion Proteins, and cell and gene therapy chapters in this book.

Disclaimer Any opinions/forward-looking statements expressed in this chapter are those of the author(s) alone and may not reflect views held by their employers (Inhibrx for Varun Ramani, Janssen Biotherapeutics for Sanjeev Bhardwaj, Zagazig University for Omnia A. Ismaiel).

References

An B, Zhang M, Johnson RW, Qu J. Surfactant-aided precipitation/on-pellet-digestion (SOD) procedure provides robust and rapid sample preparation for reproducible, accurate and sensitive LC/MS quantification of therapeutic protein in plasma and tissues. Anal Chem. 2015;87:7. https://doi.org/10.1021/acs.analchem.5b00350.

Arsene CG, Ohlendorf R, Burkitt W, Pritchard C, Henrion A, O'Connor G, Bunk DM, Güttler B. Protein quantification by isotope dilution mass spectrometry of proteolytic fragments: cleavage rate and accuracy. Anal Chem. 2008;80:11. https://doi.org/10.1021/ac7024738.

Barnidge DR, Hall GD, Stocker JL, Muddiman DC. Evaluation of a cleavable stable isotope labeled synthetic peptide for absolute protein quantification using LC-MS/MS. J Proteome Res. 2004;3:3. https://doi.org/10.1021/pr034124x.

Becher F, Ciccolini J, Imbs DC, Marin C, Fournel C, Dupuis C, Fakhry N, Pourroy B, Ghettas A, Pruvost A, Junot C, Duffaud F, Lacarelle B, Salas S. A simple and rapid LC-MS/MS method for therapeutic drug monitoring of cetuximab: a GPCO-UNICANCER proof of concept study in head-and-neck cancer patients. Sci Rep. 2017;7:1. https://doi.org/10.1038/s41598-017-02821-x.

Chen LZ, Roos D, Philip E. Development of immunocapture-LC/MS assay for simultaneous ADA isotyping and Semiquantitation. J Immunol Res. 2016;2016 https://doi.org/10.1155/2016/7682472.

Datta-Mannan A. Mechanisms influencing the pharmacokinetics and disposition of monoclonal antibodies and peptides. Drug Metab Dispos. 2019;47:10. https://doi.org/10.1124/dmd.119.086488.

De Jong KAM, Van Breugel SJ, Hillebrand MJX, Rosing H, Huitema ADR, Beijnen JH. Bottom-up sample preparation for the LC-MS/MS quantification of anti-cancer monoclonal antibodies in bio matrices. Bioanalysis. 2020;12:19. https://doi.org/10.4155/bio-2020-0204.

DelGuidice CE, Ismaiel OA, Mylott WR, Halquist MS. Optimization and method validation for the quantitative analysis of a monoclonal antibody and its related fab fragment in human plasma after intravitreal administration, using LC–MS/MS. J Chromatogr B Anal Technol Biomed Life Sci. 2021;1164 https://doi.org/10.1016/j.jchromb.2020.122474.

Devanarayan V, Smith WC, Brunelle RL, Seger ME, Krug K, Bowsher RR. Recommendations for systematic statistical computation of immunogenicity cut points. AAPS J. 2017;19:5. https://doi.org/10.1208/s12248-017-0107-3.

Dubois M, Fenaille F, Clement G, Lechmann M, Tabet JC, Ezan E, Becher F. Immunopurification and mass spectrometric quantification of the active form of a chimeric therapeutic antibody in human serum. Anal Chem. 2008;80:5. https://doi.org/10.1021/ac7021234.

El Amrani M, van den Broek MPH, Göbel C, van Maarseveen EM. Quantification of active infliximab in human serum with liquid chromatography–tandem mass spectrometry using a tumor necrosis factor alpha-based pre-analytical sample purification and a stable isotopic labeled infliximab bio-similar as internal standard. J Chromatogr A. 2016:1454. https://doi.org/10.1016/j.chroma.2016.05.070.

Elverum K, Whitman M. Delivering cellular and gene therapies to patients: solutions for realizing the potential of the next generation of medicine. Gene Ther. 2020;27(12):537–44. https://doi.org/10.1038/S41434-019-0074-7.

Faria M, Halquist MS, Yuan M, Mylott W, Jenkins RG, Karnes HT. Comparison of a stable isotope labeled (SIL) peptide and an extended SIL peptide as internal standards to track digestion variability of an unstable signature peptide during quantification of a cancer biomarker, human osteopontin, from plasma using capill. J Chromatogr B Anal Technol Biomed Life Sci. 2015;1001 https://doi.org/10.1016/j.jchromb.2015.05.040.

FDA. *Immunogenicity testing of therapeutic protein products—developing and validating assays for anti-drug antibody detection guidance for industry* 2019. https://www.fda.gov/Drugs/GuidanceComplianceRegulatoryInformation/Guidances/default.htm and/or; https://www.fda.gov/BiologicsBloodVaccines/GuidanceComplianceRegulatoryInformation/Guidances/default.htm

FDA (2018). Bioanalytical Method Validation Guidance for Industry.

Fernández Ocaña M, James IT, Kabir M, Grace C, Yuan G, Martin SW, Neubert H. Clinical pharmacokinetic assessment of an anti-MAdCAM monoclonal antibody therapeutic by LC-MS/MS. Anal Chem. 2012;84:14. https://doi.org/10.1021/ac300600f.

Findlay JWA, Smith WC, Lee JW, Nordblom GD, Das I, Desilva BS, Khan MN, Bowsher RR. Validation of immunoassays for bioanalysis: a pharmaceutical industry perspective. J Pharm Biomed Anal. 2000;21:6. https://doi.org/10.1016/S0731-7085(99)00244-7.

Furlong MT, Ouyang Z, Wu S, Tamura J, Olah T, Tymiak A, Jemal M. A universal surrogate peptide to enable LC-MS/MS bioanalysis of a diversity of human monoclonal antibody and human fc-fusion protein drug candidates in pre-clinical animal studies. Biomed Chromatogr. 2012;26:8. https://doi.org/10.1002/bmc.2759.

Furlong MT, Zhao S, Mylott W, Jenkins R, Gao M, Hegde V, Tamura J, Tymiak A, Jemal M. Dual universal peptide approach to bioanalysis of human monoclonal antibody protein drug candidates in animal studies. Bioanalysis. 2013;5:11. https://doi.org/10.4155/bio.13.55.

Halquist MS, Karnes HT. Quantification of Alefacept, an immunosuppressive fusion protein in human plasma using a protein analogue internal standard, trypsin cleaved signature peptides and liquid chromatography tandem mass spectrometry. J Chromatogr B Anal Technol Biomed Life Sci. 2011;879:11–2. https://doi.org/10.1016/j.jchromb.2011.02.034.

Ismaiel OA, Mylott WR, Jenkins RG. Do we have a mature LC-MS/MS methodology for therapeutic monoclonal antibody bioanalysis? Bioanalysis. 2017;9:17. https://doi.org/10.4155/bio-2017-4983.

Iwamoto N, Umino Y, Aoki C, Yamane N, Hamada A, Shimada T. Fully validated LCMS bioanalysis of bevacizumab in human plasma using nano-surface and molecular-orientation limited (nSMOL) proteolysis. Drug Metab Pharmacokinet. 2016;31:1. https://doi.org/10.1016/j.dmpk.2015.11.004.

Jiang H, Xu W, Titsch CA, Furlong MT, Dodge R, Voronin K, Allentoff A, Zeng J, Aubry AF, Desilva BS, Arnold ME. Innovative use of LC-MS/MS for simultaneous quantitation of neutralizing antibody, residual drug, and human immunoglobulin G in immunogenicity assay development. Anal Chem. 2014;86:5. https://doi.org/10.1021/ac5001465.

Jourdil JF, Némoz B, Gautier-Veyret E, Romero C, Stanke-Labesque F. Simultaneous quantification of adalimumab and infliximab in human plasma by liquid chromatography–tandem mass spectrometry. Ther Drug Monit. 2018;40:4. https://doi.org/10.1097/FTD.0000000000000514.

Kaur S, Liu L, Cortes DF, Shao J, Jenkins R, Mylott WR, Xu K. Validation of a biotherapeutic immunoaffinity-LC-MS/MS assay in monkey serum: "plug-and-play" across seven molecules. Bioanalysis. 2016;8:15. https://doi.org/10.4155/bio-2016-0117.

Keizer RJ, Huitema ADR, Schellens JHM, Beijnen JH. Clinical pharmacokinetics of therapeutic monoclonal antibodies. Clin Pharmacokinet. 2010;49:8. https://doi.org/10.2165/11531280-000000000-00000.

Lee JW. Generic method approaches for monoclonal antibody therapeutics analysis using both ligand binding and LC-MS/MS techniques. Bioanalysis. 2016;8:1. https://doi.org/10.4155/bio.15.231.

Lee JW, Kelley M, King LE, Yang J, Salimi-Moosavi H, Tang MT, Lu JF, Kamerud J, Ahene A, Myler H, Rogers C. Bioanalytical approaches to quantify "total" and "free" therapeutic antibodies and their targets: technical challenges and PK/PD applications over the course of drug development. In. AAPS J. 2011;13:1. https://doi.org/10.1208/s12248-011-9251-3.

Lihua H, Jerong L, Wroblewski VJ, Beals John M, Riggin RM. In vivo deamidation characterization of monoclonal antibody by LC/MS/MS. Anal Chem. 2005;77(5):1432–9. https://doi.org/10.1021/AC0494174.

Liu L. Pharmacokinetics of monoclonal antibodies and fc-fusion proteins. Protein Cell. 2018;9:1. https://doi.org/10.1007/s13238-017-0408-4.

M10 IG (2019) *Bioanalytical method validation ICH harmonized guideline M10.*

Mahomed S, Garrett N, Capparelli E, Baxter C, Zuma NY, Gengiah T, Archary D, Moore P, Samsunder N, Barouch DH, Mascola J, Ledgerwood J, Morris L, Abdool Karim S. Assessing the safety and pharmacokinetics of the monoclonal antibodies, VRC07-523LS and PGT121 in HIV negative women in South Africa: study protocol for the CAPRISA 012A randomised controlled phase I trial. BMJ Open. 2019;9:7. https://doi.org/10.1136/bmjopen-2019-030283.

European Medicines Agency. 2011. *Guideline on bioanalytical method validation*. www.ema.europa.eu/contact

Mosteller F. Data analysis and regression: a second course in statistics (Addison-Wesley Series in Behavioral Science). In *Addison-Wesley Series in Behavioral Science: Quantitative Methods* 1977.

Mou S, Huang Y, Rosenbaum A. ADME considerations and bioanalytical strategies for pharmacokinetic assessments of antibody-drug conjugates. Antibodies. 2018;7:4. https://doi.org/10.3390/antib7040041.

Mould DR, Sweeney KRD. The pharmacokinetics and pharmacodynamics of monoclonal antibodies - mechanistic modeling applied to drug development. Curr Opin Drug Discov Dev. 2007;10:1.

Neubert H, Grace C, Rumpel K, James I. Assessing immunogenicity in the presence of excess protein therapeutic using immunoprecipitation and quantitative mass spectrometry. Anal Chem. 2008;80:18. https://doi.org/10.1021/ac8005439.

Osaki F, Tabata K, Oe T. Quantitative LC/ESI-SRM/MS of antibody biopharmaceuticals: use of a homologous antibody as an internal standard and three-step method development. Anal Bioanal Chem. 2017;409:23. https://doi.org/10.1007/s00216-017-0488-2.

Ovacik M, Lin K. Tutorial on monoclonal antibody pharmacokinetics and its considerations in early development. Clin Transl Sci. 2018;11:6. https://doi.org/10.1111/cts.12567.

Roos D, Chen L, Vesapogu R, Kane C, Duggan J, Norris S. Detection of cynomolgus monkey anti-protein XYZ antibody using immunocapture-LC/MS. J Appl Bioanalysis. 2016;2:4. https://doi.org/10.17145/jab.16.016.

SA P. Correlates of protection induced by vaccination. Clin Vaccine Immunol CVI. 2010;17(7):1055–65. https://doi.org/10.1128/CVI.00131-10.

Shankar G, Devanarayan V, Amaravadi L, Barrett YC, Bowsher R, Finco-Kent D, Fiscella M, Gorovits B, Kirschner S, Moxness M, Parish T, Quarmby V, Smith H, Smith W, Zuckerman LA, Koren E. Recommendations for the validation of immunoassays used for detection of host antibodies against biotechnology products. J Pharm Biomed Anal. 2008;48(5):1267–81). Elsevier. https://doi.org/10.1016/j.jpba.2008.09.020.

Shuford CM, Sederoff RR, Chiang VL, Muddiman DC. Peptide production and decay rates affect the quantitative accuracy of protein cleavage isotope dilution mass spectrometry (PC-IDMS). Mol Cell Proteomics. 2012;11:9. https://doi.org/10.1074/mcp.O112.017145.

Sinclair A, Islam S, & Jones S (2018) Gene therapy: an overview of approved and pipeline technologies—PubMed. Sinclair A, Islam S, Jones S. Gene therapy: an overview of approved and pipeline technologies. 2018 Mar 1. In: CADTH Issues in Emerging Health Technologies. Ottawa (ON): Canadian Agency for Drugs and Technologies in Health; 2016–. 171. PMID: 30855777

Sucharski FK, Meier S, Miess C, Razavi M, Pope M, Yip R, Anderson NL, Pearson TW, Warren AP, Krantz C. Development of an automated, interference-free, 2D-LC-MS/MS assay for quantification of a therapeutic mAb in human sera. Bioanalysis. 2018;10:13. https://doi.org/10.4155/bio-2017-0252.

Talbot JJ, Calamba D, Pai M, Ma M, Thway TM. Measurement of free versus total therapeutic monoclonal antibody in pharmacokinetic assessment is modulated by affinity, incubation time, and bioanalytical platform. AAPS J. 2015;17:6. https://doi.org/10.1208/s12248-015-9807-8.

Tatarewicz S, Miller JM, Swanson SJ, Moxness MS. Rheumatoid factor interference in immunogenicity assays for human monoclonal antibody therapeutics. J Immunol Methods. 2010;357:1–2. https://doi.org/10.1016/j.jim.2010.03.012.

Vugmeyster Y. Pharmacokinetics and toxicology of therapeutic proteins: advances and challenges. World J Biol Chem. 2012;3:4. https://doi.org/10.4331/wjbc.v3.i4.73.

Wang Y, Li X, Liu YH, Richardson D, Li H, Shameem M, Yang X. Simultaneous monitoring of oxidation, deamidation, isomerization, and glycosylation of monoclonal antibodies by liquid chromatography-mass spectrometry method with ultrafast tryptic digestion. MAbs. 2016;8:8. https://doi.org/10.1080/19420862.2016.1226715.

Wu B, Chung S, Jiang XR, McNally J, Pedras-Vasconcelos J, Pillutla R, White JT, Xu Y, Gupta S. Strategies to determine assay format for the assessment of neutralizing antibody responses to biotherapeutics. AAPS J. 2016;18:6. https://doi.org/10.1208/s12248-016-9954-6.

Xu K, Liu L, Maia M, Li J, Lowe J, Song A, Kaur S. A multiplexed hybrid LC-MS/MS pharmacokinetic assay to measure two co-administered monoclonal antibodies in a clinical study. Bioanalysis. 2014;6:13. https://doi.org/10.4155/bio.14.142.

Yuan M, Ismaiel O, Mylott W. Hybrid ligand binding immunoaffinity-liquid chromatography/mass spectrometry for biotherapeutics and biomarker quantitation: how to develop a hybrid LBA-LC-MS/MS method for a protein? Rev Sep Sci. 2019;1:1. https://doi.org/10.17145/rss.19.005.

Zhou L, Hoofring SA, Wu Y, Vu T, Ma P, Swanson SJ, Chirmule N, Starcevic M. Stratification of antibody-positive subjects by antibody level reveals an impact of immunogenicity on pharmacokinetics. AAPS J. 2012;15(1):30–40. https://doi.org/10.1208/S12248-012-9408-8.

Chapter 3
An Introduction to Bioanalysis of Antibody-Drug Conjugates

Morse Faria, Varun Ramani, and Seema Kumar

Abstract Antibody-drug conjugates (ADCs) are a rapidly evolving class of biotherapeutics mostly developed for the treatment of cancer. The ADC molecule typically comprises of a cytotoxic small-molecule drug covalently bound to a monoclonal antibody via a linker. Because of their complex structure combining large- and small-molecule drug characteristics and their heterogeneous and dynamic nature, unique bioanalytical strategies are needed to identify, characterize, and quantify ADCs for their safety and efficacy assessments. ADC bioanalysis thus necessitates an integrated bioanalytical approach including both ligand-binding assays and liquid chromatography coupled with mass spectrometry-based assays, conventionally used for large- and small-molecule bioanalysis, respectively. This chapter provides an introductory understanding of the structure and chemistry of ADC molecules and various bioanalytical strategies used for their pharmacokinetic and immunogenicity assessment.

Keywords Antibody-drug conjugates (ADC) · Total antibody · Conjugated payload · Conjugated antibody · Unconjugated payload · Hybrid LC-MS/MS · DAR characterization

3.1 Introduction

Antibody-drug conjugates (ADCs) typically consist of small-molecule drug (aka payload) covalently linked to recombinant monoclonal antibodies (mAbs) via synthetic linkers. By combining the high specificity and longer half-life of mAbs with the high potency of small-molecule drugs, ADCs aim to selectively deliver highly

M. Faria (✉)
Clinical Research Group, Thermo Fisher Scientific, Richmond, VA, USA

V. Ramani
Inhibrx, La Jolla, CA, USA

S. Kumar
EMD Serono, Inc. (a business of Merck KGaA, Darmstadt, Germany), Billerica, MA, USA

S. Kumar (ed.), *An Introduction to Bioanalysis of Biopharmaceuticals*, AAPS Advances in the Pharmaceutical Sciences Series 57,
https://doi.org/10.1007/978-3-030-97193-9_3

cytotoxic drugs to targeted tissue, thereby limiting systemic exposure and increasing therapeutic index of small-molecule drugs. Majority of the ADCs currently in development are for oncology indications. The first ADC to gain regulatory approval from the US Food and Drug Administration (US FDA) was gemtuzumab ozogamicin in 2000, commercially known as Mylotarg®. In the last decade, nine additional ADCs have been approved by US FDA (see Table 3.1), and more than 50 ADCs against diverse tumor targets are in various phases of clinical development.

Albeit their initial clinical success, the clinical development of ADCs is still hindered by their relatively narrow therapeutic index. Although tumor-specific antibodies allow enrichment of cytotoxic payloads in tumors, adverse safety events frequently occur before ADCs reached their optimal efficacious dose, thereby limiting their clinical response. To overcome some of these limitations, advancements have been made both in target selection and in design and development of ADCs such as, type of payload and linker, conjugation chemistry, number of payloads conjugated to an antibody (Ab) molecule, location of the conjugation site(s), and the antibody framework (Beck et al. 2017; Perez et al. 2014).

The multi-component structure of ADC requires a combination of bioanalytical strategies typically applied for small-molecule drugs and large-molecule biotherapeutics. Additionally, safety and efficacy assessment of ADC requires multiple bioanalytical assays to monitor its catabolic fate in vivo (pharmacokinetics [PK], immunogenicity, biotransformation, etc.). In this chapter, we will discuss various bioanalytical strategies to support PK assessment and immunogenicity evaluation of ADC molecules.

3.2 Chemistry of Antibody-Drug Conjugates

The success of an ADC depends on the careful selection of its building blocks (mAb, payload and linker) and the conjugation chemistry employed to bring together these building blocks for ADC synthesis.

3.2.1 Antibody

The mAb component of ADC is usually designed to target antigens that are highly expressed on tumor cells relative to normal healthy cells, to achieve high tumor efficacy while maintaining low systemic toxicity. The advancement of protein engineering technologies has led to a new generation of mAb-based protein scaffolds for ADC synthesis including bispecific antibodies, biparatopic antibodies, and Probody™ therapeutics (Beck et al. 2017; Perez et al. 2014; Polu and Lowman 2014; Faria et al. 2019a).

Table 3.1 List of approved ADC molecules

ADC	Commercial name	Indication	Target antigen	Antibody	Payload	Linker	Average DAR	FDA approval
Gemtuzumab ozogamicin	Mylotarg®	Relapsed or refractory acute myeloid leukemia	CD33	Humanized IgG4	Calicheamicin	Hydrazone	2.5	2000, withdrawn in 2010 and reapproved in 2017
Brentuximab vedotin	Adcetris®	Relapsed or refractory Hodgkin lymphoma and systemic anaplastic large cell lymphoma	CD30	Chimeric IgG1	MMAE	Valine-citrulline	4	2011
Trastuzumab emtansine	Kadcyla®	HER2-positive metastatic breast cancer	HER2	Humanized IgG1	DM1	SMCC	3.5	2013
Inotuzumab ozogamicin	Besponsa®	Relapsed or refractory acute lymphoblastic leukemia	CD22	Humanized IgG4	Calicheamicin	Hydrazone	6	2017
Polatuzumab vedotin-piiq	Polivy®	Diffuse large B-cell lymphoma	CD79b	Humanized IgG1	MMAE	Valine-citrulline	4	2019
Trastuzumab deruxtecan	Enhertu®	HER2-positive metastatic breast cancer and locally advanced or metastatic gastric cancer	HER2	Humanized IgG1	DXd	Glycine-glycine-phenylalanine-glycine (cleavable)	7.5	2019
Enfortumab vedotin	Padcev®	Locally advanced or metastatic urothelial cancer	Nectin-4	Fully human IgG1	MMAE	Valine-citrulline	3.8	2019
Sacituzumab govitecan	Trodelvy®	Metastatic triple-negative breast cancer	Trop-2	Humanized IgG1	SN-38	Carbonate	7.6	2020
Belantamab mafodotin	Blenrep®	Relapsed or refractory multiple myeloma	BCMA	Humanized IgG1	MMAF	MC	4	2020
Loncastuximab tesirine	Zynlonta®	Relapsed or refractory large B-cell lymphoma	CD19	Humanized IgG1	SG3199 (PBD dimer toxin)	Valine-alanine	2	2021

BCMA B-cell maturation antigen, *DM1* mertansine; *DXd* deruxtecan, *HER2* human epidermal growth factor receptor 2, *MC* maleimidocaproyl, *MMAE* monomethyl auristatin E, *MMAF* monomethyl auristatin F, *SN-38* 7-ethyl-10-hydroxycamptothecin, *Trop2* trophoblast cell surface antigen 2, *SMCC* succinimidyl 4-(N-maleimidomethyl)cyclohexane-1-carboxylate

3.2.2 *Payload*

The majority of payloads currently used in ADC development fall into three main categories depending on their mechanism of action (MoA)—anti-mitotic (tubulin filaments damaging), deoxyribonucleic acid (DNA) damaging, and transcription inhibitors. Table 3.2 lists the major classes of payloads and their MoA.

3.2.3 *Linker*

ADC linkers are the essential component of an ADC molecule that holds the antibody and payload together. Selection of an appropriate linker is of a key importance in the construction of an ADC molecule. Linker stability during storage and systemic circulation, and to efficiently release payload in a selective environment are the ideal characteristics of ADC linkers. Linker chemistry is thus an important consideration in ADC bioanalysis.

ADC linkers can be classified into two major classes—cleavable linkers and non-cleavable linkers (Perez et al. 2014; Tsuchikama and An 2018). Majority of the approved ADCs have cleavable linker chemistry.

3.2.3.1 Cleavable Linker

Cleavable linkers are designed to selectively release payload from ADC in target cells either by responding to an environmental difference (e.g., pH, redox potential, etc.) between the extracellular and intracellular environments or by cleavage from specific lysosomal enzymes (Cathepsin B, glucuronidase, etc.) in target cells. Adcetris® (Brentuximab vedotin) is an example of an ADC having a Cathepsin B

Table 3.2 List of major payload classes

Sr. no.	Payload class	Examples of payloads	Mechanism of action (MoA)
1	Auristatins	Monomethyl Auristatin E (MMAE) and Monomethyl Auristatin F (MMAF)	Anti-mitotic
2	Calicheamicin	N-Acetyl-γ-calicheamicin	DNA damaging
3	Maytansinoids	Mertansine/Emtansine (DM1) and Ravtansine/Soravtansine (DM4)	Anti-mitotic
4	Pyrrolobenzodiazepines (PBDs)	SG3199	DNA damaging
5	Tubulysins	AZ13599185	Anti-mitotic
6	Duocarmycin	BMS-936561 and MED-2460	DNA damaging
7	Camptothecin analogues	SN-38 and DX-8951f	DNA topoisomerase 1 inhibition

cleavable linker comprising of valine-citrulline (Val-Cit) dipeptide bound to p-aminobenzyloxycarbonyl (PABC).

3.2.3.2 Non-cleavable Linker

ADCs with non-cleavable linkers are designed to undergo complete degradation of the Ab by cytosolic and lysosomal proteases after internalization into target cells, resulting in release of free payload along with the linker attached to a portion of amino acid residues from the Ab. Kadcyla®, a humanized anti-human epidermal growth factor receptor 2 (HER2) antibody-maytansine conjugate, aka Trastuzumab emtansine (T-DM1), is an example of an ADC where non-cleavable linkers have been successfully employed (Erickson et al. 2010).

3.2.4 ADC Conjugation Chemistry

Various conjugation chemistries have been employed for ADCs—ranging from conventional lysine-based conjugation (employed in Kadcyla™, Mylotarg™, and Besponsa™) or cysteine-based conjugation (employed in Adcetris™, Polivy™, Padcev™, Enhertu®, Blenrep) to site-specific conjugations.

Depending on the conjugation chemistry employed, the conjugation process can result in a mixture of ADCs species with variable drug-to-antibody ratio (DAR) and different locations of conjugation sites. For instance, conventional lysine- and cysteine-based random conjugation chemistries typically yield a heterogeneous mixture of DAR species. But the novel site-specific conjugation chemistries (including the use of engineered cysteine residues, unnatural amino acids, enzymatic conjugation through glycotransferases and transglutaminases) have been shown to yield homogeneous composition of ADC.

The average DAR value of ADC could also change once ADC enters the systemic circulation irrespective of its DAR in the starting material. This change in DAR can impact ADC disposition and in turn can impact its safety and efficacy profile. ADCs with varying DAR require additional bioanalytical considerations to ensure comprehensive PK characterization of its various DAR species.

3.3 ADC Mechanism of Action

ADCs are designed to kill tumor cells in a target-dependent manner. The key steps involved in MoA of ADC includes (1) binding of ADC to its specific tumor antigen on the target cells, (2) internalization of the tumor antigen-ADC complex via receptor-mediated endocytosis, (3) trafficking from endosomal vesicles to the lysosomes, (4) intracellular release of the cytotoxic payload in the acidic and proteolytic rich environment of lysosomes, and (5) cell death by the released payload. The mechanism of cell death can vary based on the type of payload used (e.g., disruption of cytokinesis by tubulin polymerization inhibitors such as maytansines and auristatins, DNA damage by DNA interacting agents such as calicheamicins, duocarmycins, pyrrolobenzodiazepines, etc.). Neighboring tumor antigen-positive as well as tumor antigen-negative cancer cells may also be killed when free drug is released into the tumor environment by the dying cell in a process known as the bystander effect (Staudacher and Brown 2017).

3.4 PK Considerations for Pre-Clinical and Clinical Development of ADC

The PK of ADC is more complex than a typical biotherapeutic due to the presence of multiple functional components (mAb, payload, and linker) and the complex interplay among these components. Though all three components of ADC contribute to its in vitro and in vivo stability, the overall PK of ADC is usually dominated by its mAb component (Lin and Tibbitts 2012). Similar to mAb biotherapeutics, the PK profile of an ADC is characterized by low clearance rate, long circulating half-life (days), limited volume of distribution, and targeted tissue distribution. Though as such, the cytotoxic payload does not drive PK of an ADC, the payload properties such as its MoA, potency (typically in nanomolar range), hydrophobicity, permeability, metabolism, efflux transporter profiles, etc. can impact the PK of an ADC. The differences in payload MoA (DNA damaging vs. microtubule inhibitors) in particular could play a role in changing the PK driver (such as C_{max}, AUC, or time above a concentration threshold) of efficacy and/or toxicity of ADC (Kamath and Iyer 2016).

Other factors such as ADC physicochemical attributes including DAR, conjugation chemistry, and site of conjugation can also influence the PK and disposition of ADC. The stability of a linker has a significant impact on ADC PK. Irrespective of its chemistry (cleavable vs non-cleavable), a linker should remain stable in circulation to minimize systemic toxicity but efficiently release the active payload once it is internalized by the target tumor cell. However, even a stable linker could undergo deconjugation in systemic circulation (Perez et al. 2014). Both ADC and its payload can undergo in vivo biotransformation that may have an impact on their safety and/or efficacy profile (Shen et al. 2012; Saad et al. 2015; Lee et al. 2020). The ADC biotransformation assessments are typically carried out using in vitro plasma stability and lysosomal stability studies. These studies also aid in the identification of ADC catabolites and payload metabolites that guides the bioanalytical (BA) strategy for ADC quantitation.

Other factors impacting the PK of ADC include soluble or shed target (Wang et al. 2016; Kaur et al. 2013) and endogenous anti-drug antibodies that may form complexes with ADC (Saad et al. 2015). ADCs with thiol-maleimide linker chemistry may also undergo linker-payload deconjugation through a thiol exchange process primarily with reactive cysteines on the plasma albumin resulting in arming these endogenous proteins with potent payload (Beck et al. 2017; Rago et al. 2017; Dong et al. 2018; Wei et al. 2016). These albumin-payload adducts can give rise to immunotoxicity and hence, need to be monitored.

3.5 Bioanalysis of ADC to Support their PK Assessment

Typically, the bioanalysis of a biotherapeutic involves measurement of its concentration (and its major metabolite, if needed) over time in the biological matrix (serum, plasma, urine, tissues, etc.) to determine its PK profile. However, ADC bioanalysis is challenging due to its multiple functional components and its dynamic and heterogenous nature. The heterogeneity in starting reference material may evolve further in vivo due to biotransformation, spontaneous or environment induced deconjugation of payload from the ADC by enzymatic or chemical reactions, and differences in the clearance rate of various DAR species (DiJoseph et al. 2004; Junutula et al. 2008). Figure 3.1 illustrates the complexity of ADC species in vivo. Thus, multiple ADC analytes are typically utilized to determine PK and overall disposition of ADC in vivo.

The commonly used ADC analytes include total antibody (TAb), total ADC, unconjugated payload, changes in DAR distribution over time, and anti-drug antibodies (ADAs) against any functional domain of ADC molecule. These analytes are summarized in Table 3.3. Each analyte provides a unique information of ADC behavior in vivo, and there is no single bioanalytical assay that can quantitate all ADC analytes.

- Total antibody analyte represents the total antibody (Ab) component of an ADC, including both conjugated and unconjugated forms of ADC (DAR $\geq$ 0). The purpose of this analyte is to determine if PK characteristics of the Ab component

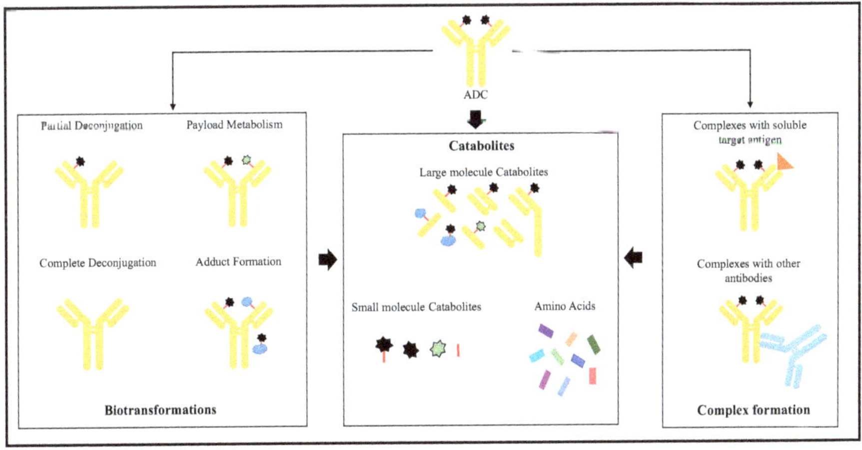

Fig. 3.1 ADC complexity due to its in vivo biotransformation and catabolism

Table 3.3 Commonly measured analytes in ADC bioanalysis

Analyte Type	Description	Typical analytical platforms
Total antibody	Total antibody (conjugated or unconjugated)	LBA, hybrid LC-MS/MS
Total ADC	Total conjugated antibody (DAR > 1) OR	LBA
	Total payload conjugated to antibody (DAR > 1)	LBA, hybrid LC-MS/MS
Unconjugated payload	Payload not conjugated to antibody	LC-MS/MS
Anti-drug antibodies (ADA)	Antibodies directed against antibody components of ADC, linker, or drug (binding/neutralizing)	LBA

ADC Antibody-drug conjugate, *DAR* drug-antibody ratio, *LBA* ligand binding assay, *LC-MS/MS* liquid chromatography and tandem mass spectrometry

of ADC is indeed representing characteristics typically expected from an Ab, and is not impacted by conjugation of linker-payload.

- Total ADC (DAR > 0) analyte represents the intact conjugated ADC, which is the active therapeutic analyte. It can be quantitated either as conjugated-payload or conjugated-antibody analytes.
- Unconjugated payload represents payload present in circulation that is no longer covalently bound to the ADC. Comparison of the TAb and the total ADC PK profiles can provide an understanding of the in vivo deconjugation or the changes in the DAR distribution of ADC molecules.

3.5.1 *Bioanalytical Platforms for ADC*

As the ADC analytes comprise of both large and small molecules, multiple bioanalytical assays and platforms are needed to analyze diversity of ADC analytes. While traditional small-molecule assays using liquid chromatography separation with mass spectrometry detection (LC-MS/MS) are commonly used for unconjugated payload and DAR distribution over time, both ligand binding assay (LBA) and hybrid LC-MS/MS could be used for TAb and total ADC analytes. The choice of analytical platform depends on the availability of critical reagents, desired assay sensitivity, and the PK question that needs to be addressed at the respective stage of drug development (Kaur et al. 2013; Mou et al. 2018; Cahuzac and Devel 2020).

3.5.1.1 Ligand Binding Assays for ADC Bioanalysis

Ligand binding assays are the gold standard for bioanalysis of large molecules. LBAs are routinely used to measure TAb and total ADC (via conjugated-antibody assay) exposure by taking advantage of antibodies specific to the mAb or payload component of ADC, respectively. Figure 3.2 illustrates the LBA assay design for the measurement of TAb and total ADC (conjugated-antibody) analytes.

Various LBA-based bioanalytical platforms including conventional enzyme-linked immunoassay (ELISA), mesoscale discovery (MSD), and Gyros could be employed for ADC bioanalysis. Each platform offers unique advantages and limitations for LBA applications. For example, MSD and Gyros platform claim to offer higher sensitivity, broad dynamic range, low sample volume, and/or reduced matrix interference compared to conventional ELISA platform. (Wang et al. 2016; Kaur et al. 2013; Mou et al. 2018). LBA platform allows development of high-throughput, cost-effective, and easy to implement assays for ADC bioanalysis. Figure 3.2 illustrates an ELISA platform design for the measurement of total antibody and total ADC (conjugated-antibody).

Total Antibody by LBA

Total antibody assay monitors Ab concentration regardless of whether payload is attached to the Ab or not, i.e., fully conjugated, partially deconjugated, and fully deconjugated ADC (Fig. 3.2a). This assay is used for overall PK assessment of the ADC (Gorovits et al. 2013; Stephan et al. 2011). The reagents typically employed for capture and detection in TAb assay include specific reagents such as recombinant target protein and anti-idiotypic mAbs or generic reagents such as antibodies directed against the whole human immunoglobulin G (IgG) framework or against the (Fab')2 region, or against the Fc region, or against the light chain (LC), or against the heavy plus light (H + L)-chain regions. These reagents bind to the antibody component of ADC regardless of its DAR value. Though these reagents do not directly bind to the payload, due to steric hindrance, payload can indirectly influence binding of these reagents to the Ab component of ADC. This interference might be more prominent for higher DAR species. As a result, the assay may not accurately estimate all expected DAR species in systemic circulation, thereby affecting the observed overall PK characteristics of the ADC.

Total ADC (Conjugated Antibody) by LBA

Conjugated-antibody assay monitors concentration of Ab that bears at least one payload molecule (i.e., ADC with DAR $\geq$ 1) (Fig. 3.2b). The critical reagents typically employed in the conjugated Ab assay bind both the payload component and the Ab component of ADC. Similar to the TAb assay, conjugated Ab assays also exhibit sensitivity to the site of conjugation and the DAR heterogeneity of ADC. The binding of anti-payload antibodies to the payload component of ADC might be hindered by the solvent accessibility of the conjugation site. In addition, proportional binding of reagents to the Ab component of ADC may not be possible due to the steric hindrance by the multiple adjacently conjugated payload molecules. Thus, the conjugated-antibody assay may also provide inaccurate measurement of higher and lower DAR species in systemic circulation.

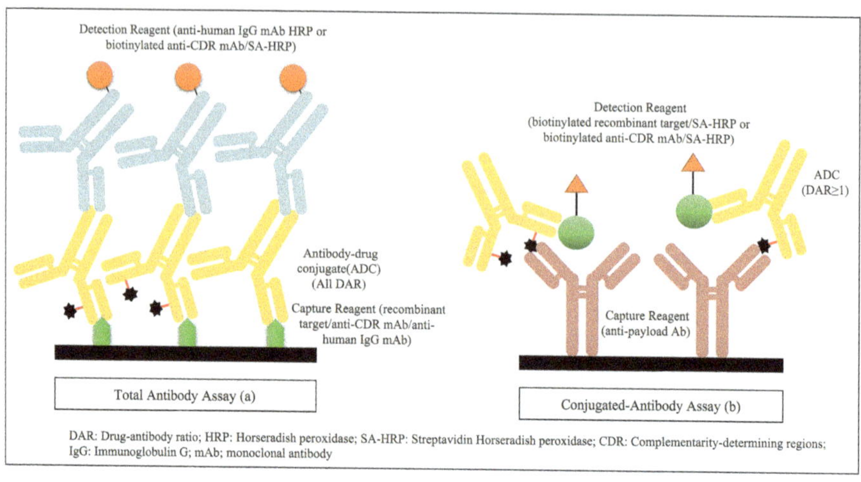

Fig. 3.2 ELISA assays for ADC bioanalysis

It is recommended that LBAs designed to measure TAb and total ADC (conjugated Ab) analytes should be assessed for DAR sensitivity using individual purified or enriched DAR species (Kaur et al. 2013). Since the LBA sensitivity to DAR depends on the linker type, payload class, conjugation site, and conjugation chemistry, it may be necessary to screen multiple formats and reagents in relevant biological matrices to achieve desired DAR sensitivity or insensitivity (Stephan et al. 2008).

Similar to LBA for biologics, TAb and total ADC (conjugated Ab) assay performance are also impacted by shed or soluble target and/or ADA as they may block the capture and/or detection reagent binding to the ADC, particularly when specific reagents such as anti-idiotypic Ab or recombinant target antigens are used as reagents. In non-clinical samples, the shed or soluble target may not be cross-reactive to ADC, or the soluble or shed target level may be too low to have a significant impact on the assay performance (Wang et al. 2016). However, in clinical samples, the levels of shed or soluble target depending on the patient population may be high and/or variable. They could thus significantly impact TAb and total ADC (conjugated Ab) quantitation in clinical studies (Wang et al. 2016). The presence of high levels of shed or soluble targets is an important bioanalytical consideration during the development of a TAb assay for clinical applications.

3.5.1.2 Mass Spectrometry for ADC Bioanalysis

It is challenging to develop LBA for small molecules due to the lack of tertiary structure in small-molecule drugs and potential steric hindrance in binding of capture and detection reagents to small-molecule drugs. LC-MS/MS is thus the platform of choice for bioanalysis of unconjugated payload and its metabolite. In the last decade, hybrid LC-MS/MS assays have emerged as alternatives to LBA for bioanalytical measurement of TAb and conjugated payload of ADCs. Hybrid LC-MS/MS combines the high selectivity of LBA with the selectivity and sensitivity of an LC-MS/MS method. The development of critical reagents for LBA often involves resource-intensive and time-consuming processes that restrict their availability during the early stages of drug development. Hybrid LC-MS/MS assays offer rapid method development alternatives to LBA at these early stages of development. Additionally, hybrid LC-MS/MS overcomes the limitations of LBA for providing information about the drug load or DAR of ADC.

High-resolution mass spectrometry (HRMS) coupled with immunoaffinity approach is employed for monitoring in vivo ADC biotransformation and DAR characterization using intact protein analysis (Huang et al. 2020).

Hybrid LC-MS/MS

Typical, mainstream hybrid LC-MS/MS ADC assays utilize capture reagents targeted toward the Ab component of ADC. The surrogate analyte is then released using the chemical or enzymatic cleavage of ADC. In TAb assay, the surrogate analyte is a unique peptide (aka signature peptide) from the antibody component, and in conjugated-payload assay, the surrogate analyte is the released payload or payload attached to the linker or linker-peptide.

Hybrid LC- MS/MS assay consists of three major steps:

(a) Immunocapture

This step is essential for the isolation and enrichment of ADC molecules from the complex biological matrix. The capture reagent is selected based on the complexity of the matrix (such as serum, plasma, cerebrospinal fluid), species (animal or human), target concentration range, availability of capture reagent, and analytical question (target-free or target-bound ADC). Similar to LBA, the capture reagent could be a generic capture reagent (e.g., anti-human IgG, anti-human Fc, Protein A or G, etc.) or a specific capture reagent (e.g., recombinant target protein, anti-idiotypes against Ab component of ADC, or Ab against payload component of ADC, etc.).

The immunocapture involves a stepwise approach- -immobilization of the biotinylated capture reagent on streptavidin-coated magnetic beads followed by the addition of ADC sample to allow ADC binding to the immobilized capture reagent. The beads containing immunocaptured ADC are then washed using a plate washer with magnets or automated magnetic bead transfer systems, to remove residual matrix components attached to the beads.

The magnetic beads allow higher capture capacity than microtiter plates, thereby provide a broader dynamic range of the assay. Immunocapture of an ADC can be performed in a stepwise manner as indicated above, or it can also be performed in a homogeneous fashion where ADC samples are pre-incubated with biotinylated capture reagent.

(b) Release of Surrogate Analyte

Once the ADC molecule is immunocaptured, the surrogate analyte is released. Depending on the type of ADC analyte, the surrogate analyte could be a peptide or a small-molecule drug. For instance, in TAb assay, the Ab component of ADC is enzymatically proteolyzed, typically using trypsin, to yield a signature peptide that is used as a surrogate analyte.

For ADCs with cleavable linkers, the payload or payload-linker released from ADC is used as a surrogate analyte in the conjugated payload assay. For ADCs with non-cleavable linkers, the antibody component of ADC is completely proteolyzed to yield a surrogate analyte comprising of the peptide-linker-payload.

(c) Detection

Detection of the surrogate analyte is done by LC-MS/MS. An appropriate internal standard (IS) is added depending on the assay workflow (see Fig. 3.4). Additional clean-up steps such as protein precipitation or solid-phase extraction (for conjugated payload assay) or peptide immunoaffinity isolation (for TAb assay) steps may be required to improve selectivity or to achieve higher sensitivity. High-resolution mass spectrometry is utilized for DAR characterization or for monitoring the biotransformation using intact MS analysis (Xu et al. 2013; Jin et al. 2018; Huang et al. 2021). Hybrid LC-MS/MS allows multiplexing of multiple ADC assays. The TAb and conjugated payload analytes were simultaneously monitored in a single assay for bioanalysis of ADC with cleavable dipeptide linker (Faria et al. 2019a). Similarly, multiple peptides from different domains of the ADC antibody may be monitored to ensure integrity of the biotherapeutic while in systemic circulation (Huang et al. 2020).

Hybrid-LC-MS/MS for Total Antibody Assay Quantitation

Traditionally, LBA has been the preferred platform for measurement of TAb analyte (see Sect. 3.5.1.1). Lately, hybrid LC-MS/MS has emerged as an alternate platform for TAb measurement in nonclinical samples (Kaur et al. 2013; van den Broek et al. 2013; Furlong et al. 2014). In this assay format, the ADC is isolated using immunoaffinity approaches from the matrix using generic capture reagent such as Protein A beads or polyclonal anti-human antibodies followed by enzymatic proteolysis, typically with trypsin, to generate a signature peptide from the Fc region of the antibody. The ease of method development of TAb assays using hybrid LC-MS/MS has been demonstrated by the "plug and play" approach (Kaur et al. 2016). Commercially available stable labeled IgG antibodies, such as SILu™MAb, are typically used as internal standards for this assay design (Faria and Halquist 2018; Moucun et al. 2019). See Fig. 3.3 for a typical workflow of hybrid LC-MS/MS method for quantification of TAb analyte for ADC.

Similar to LBA, generic capture reagents cannot be used in hybrid LC-MS/MS assays for clinical applications. This is due to the presence of excessive amounts of endogenous human IgGs in clinical samples that could interfere in the assay. Thus,

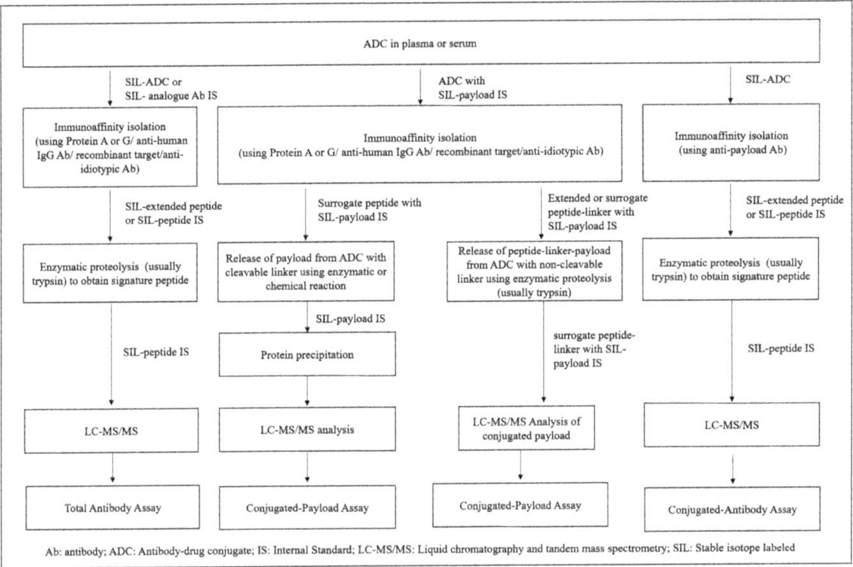

Fig. 3.3 Workflow of hybrid LC-MS/MS assays for ADC bioanalysis

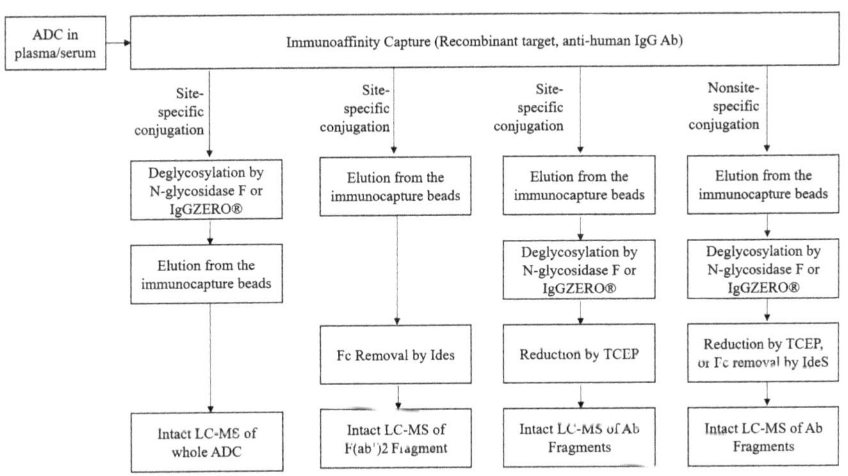

Fig. 3.4 Workflow for DAR characterization

specific capture reagents such as recombinant target antigens, anti-idiotypic or anti-complementarity-determining region (CDR) antibodies, etc. are used for immunocapture step in clinical hybrid LC-MS/MS assays (Kaur et al. 2013; Huang et al. 2020; Faria et al. 2019b). The signature peptide from the CDR region of Ab is used as a surrogate analyte for the clinical hybrid LC-MS/MS assays. Generating a stable isotope labeled ADC to serve as an internal standard in the hybrid LC-MS/MS assays is often challenging. Thus, typically stable isotope labeled (SIL) peptides are used as internal standards and are added to the test sample after the immunocapture step. The lack of IS to monitor the immunocapture step of ADC is a limitation of this assay design. However, robust assay performance has been demonstrated for hybrid LC-MS/MS assays where no IS was added in the immunocapture step (Wang et al. 2016).

Similar to LBA, soluble or shed target interference could also pose a challenge for hybrid LC-MS/MS-based TAb assay. Soluble or shed circulating targets may block the binding of specific reagents (anti-idiotypic Ab or recombinant target antigens) to ADC and thus interfere in the immunoaffinity capture step. In vivo biotransformation of the Ab component of ADC is another challenge for hybrid LC-MS/MS-based TAb assessment. For instance, deamidation of asparagine and isomerization of aspartic acid are two common biotransformation of Ab that need to be considered for choosing the signature peptide (Wei et al. 2018).

Since hybrid LC-MS/MS platform allows for the monitoring of several peptides simultaneously, both naïve and modified forms of the peptide can be monitored. In addition, multiple peptides from different subdomains of the antibody can be monitored. Collectively, these data can provide a complete picture of biotransformation of the Ab component of ADC.

Similar to LBA-based TAb assay, another consideration during the development of hybrid LC-MS/MS-based TAb assay is that there should be no assay bias due to changes in ADC DAR in vivo. For example, higher DAR ADC may be underestimated in the assay due to the steric hindrance from payload that may prevent binding of the ADC to the capture reagent. Hence, it may be necessary to test enriched ADC fractions with varying DARs for assay recovery to ensure that there is no capture bias during assay development.

Irrespective of the analytical platform (LBA or hybrid LC-MS/MS) used for TAb assay, during assay development, the assay is evaluated for performance characteristics including but not limited to accuracy, precision, dilutional linearity, and specificity. A linear regression model is used for quantitation for the hybrid LC-MS/MS assays. Hence, an immunocapture capacity evaluation is performed to ensure the immunoaffinity isolation step has adequate capacity to distinguish between the upper limit of quantification (ULOQ) and samples above ULOQ concentrations. To evaluate immunocapture capacity, quality control (QC) samples above ULOQ concentration are analyzed undiluted. The immunocapture capacity is measured as the extrapolated concentration of the undiluted QC sample above ULOQ. An immunocapture capture capacity of twice the ULOQ concentration is considered adequate for a hybrid LC-MS/MS method.

Hybrid-LC-MS/MS for Total ADC (Conjugated-Antibody) Quantitation

Conventionally, the conjugated-Ab analyte is quantitated using LBA (see Sect. 3.5.1.1). But hybrid LC-MS/MS analytical platforms can also be employed for measuring conjugated antibody (see Fig. 3.3). In this approach, the ADC is captured using anti-payload antibody followed by its enzymatic digestion and LC-MS/MS detection using signature peptide using obtained after. Similar to hybrid LC-MS/MS-based TAb analysis, a peptide from the human Fc region or from the CDR region is used as the signature peptide for nonclinical studies, while for clinical studies, a signature peptide from the CDR region is used.

Hybrid-LC-MS/MS for Total ADC (Conjugated-Payload) Quantitation

The conjugated-payload assay is designed to measure the concentration of payload molecules that are conjugated to the antibody. The conjugated payload involves direct measurement of the potent bioactive payload and hence may provide a more accurate assessment of the payload exposure to the target site and in turn may provide higher exposure vs. efficacy correlation (Kumar et al. 2015).

Hybrid LC-MS/MS is the main platform for conjugated payload measurement. In the hybrid LC-MS/MS-based conjugated-payload assay, the capture reagent is directed towards the Ab component of the ADC molecule. So, the immunocapture step is similar to hybrid LC-MS/MS-based TAb assay. Once isolated using the immunoaffinity approach, the payload molecule is released from the ADC, and the cleaved payload is measured by LC-MS/MS. Depending on the linker chemistry, an enzymatic or chemical reaction is employed for cleaving the payload from ADC. Though the surrogate analyte in the conjugated payload assay is a small-molecule drug, the use of hybrid immunocapture techniques warrants the large-molecule bioanalytical method validation guidelines for this assay (Lee et al. 2020; Kaur et al. 2013; Gorovits et al. 2013).

Hybrid LC-MS/MS offers an advantage for monitoring ADC biotransformation. For instance, the payload and/or the linker may undergo metabolism while it still remained attached to the Ab in systemic circulation. MEDI4276 is an ADC comprising of a humanized antibody attached via a protease-cleavable peptide-based maleimidocaproyl linker to a tubulysin analogue (AZ13599185) (Faria et al. 2019b). The tubulysin analogue is known to undergo deacetylation without deconjugation from the antibody in vivo. To support the clinical studies of MEDI4276, two hybrid LC-MS/MS assays were employed to measure conjugated-payload and the conjugated-deacetylated payload. A modified form of the ADC was synthesized with deacetylated payload and used as reference standard for the conjugated-deacylated payload assay.

Conjugated-payload quantification by hybrid LC-MS/MS has its limitations. The major limitation is in the analysis of ADCs with non-cleavable linkers, wherein after immunoaffinity isolation of ADC, the Ab component of ADC needs to be completely proteolyzed to release peptide-linker-conjugated-payload for the surrogate analyte. In these cases, the reference standard and the stable labeled IS may not be readily available. Additionally, non-cleavable, lysine-based random conjugated ADCs offer challenges as they form multiple peptide fragments linked to the conjugated payload after complete proteolysis of the ADC molecule. The immunocapture step for conjugated payload assay faces similar challenges as hybrid LC-MS/MS-based TAb assay and hence needs to be appropriately mitigated (Sect. 4.5.2.1.1).

LC-MS/MS-Based ADC Bioanalysis

Unconjugated payload assay involves measurement of payload deconjugated from ADC in plasma and/or in target tissues after ADC administration and the trace amount of unconjugated payload present in the ADC formulation. LC-MS/MS is the platform of choice for bioanalysis of unconjugated payload, though LBA has also been used depending on the availability of the anti-payload antibody (Buckwalter et al. 2004). Unconjugated payload bioanalysis is similar to small-molecule bioanalysis using LC-MS/MS. The sample extraction is carried out using techniques such as protein precipitation (Huang et al. 2020; Faria et al. 2019b; Grafmuller et al. 2016), solid-phase extraction (Heudi et al. 2016), and supported liquid extraction (Wei et al. 2012).

Unconjugated Payload Quantitation

The exposure-response correlation for ADC depends on its in vivo stability and its ability to efficiently release payload in target tumor cells. The unconjugated payload assay provides an understanding of the ADC stability in systemic circulation and in target tumor microenvironment.

The bioanalytical strategy used for unconjugated payload assay depends on its release mechanism of ADC, its metabolism, and the nature of its major metabolites. For instance, while the deconjugation of ADC with cleavable linker releases the free payload, the deconjugation of ADC with non-cleavable linker can produce a range of ADC catabolites such as amino acid-linker-payload and linker-payload moieties in addition to free payload (Jain et al. 2015). The unconjugated payload analysis for Kadcyla® included measurement of DM1, Lys-mcc-DM1, and mcc-DM1 (Shen et al. 2012; Kaur et al. 2013).

The major challenge of unconjugated payload assay is the assay sensitivity. Due to higher cytotoxic activity of payload, ADCs are typically dosed at lower concentrations making it difficult to quantitate lower levels of deconjugated payload (Heudi et al. 2016; Wei et al. 2012). Additionally, stable ADCs release relatively lower amounts of free payload in systemic circulation. Furthermore, to reduce metabolic liability, ADC payloads are often more hydrophobic and have higher molecular weight (>700 Da) in comparison to conventional small molecules that adds to the analytical challenges.

Another challenge for unconjugated payload quantification is that the analyte sometimes need to be stabilized prior to the LC-MS/MS analysis. For example, in the study described by Heudi et al., the DM1 released from thioether-linked trastuzumab-DM1 contains a free thiol moiety, which needed to be reduced and alkylated prior to its analysis. These steps prevent free DM1 forming a dimer or reacting with the free thiol group on endogenous compounds such as cysteine or glutathione (Heudi et al. 2016). Similarly, in the quantification of unconjugated DM4 described by Wei et al., the unconjugated DM4 was stabilized using a LLE procedure in an ice water bath (Wei et al. 2012). Similarly, with unconjugated payload assays for ADCs with acid-labile hydrazone linkers, the pH adjustment is needed, and protease inhibitor cocktails are added to inhibit ex vivo cleavage of enzyme-cleavable linkers (Wei et al. 2018).

The deconjugated payload for maleimide-containing ADCs can form adducts with endogenous proteins leading to underestimation of unconjugated payload concentration in test samples (Alley et al. 2008). The adduct formation can be assessed using hybrid LC-MS/MS (Dong et al. 2018; Grafmuller et al. 2016). Furthermore, to evaluate the total deconjugated payload, additional steps may be incorporated in the bioanalytical workflow to release the payload bound to endogenous proteins prior to its measurement (Wei et al. 2018).

ADC payload can undergo biotransformation in the conjugated and unconjugated form, and thus biotransformation of both forms may need to be monitored. MEDI4276 is an ADC containing a tubulysin toxin (AZ13599185) that is known to undergo in vivo deacetylation. Faria et al. reported a simultaneous LC-MS/MS assay for monitoring both unconjugated payload and its unconjugated deacetylated form (Faria et al. 2019b).

The unconjugated payload assays are validated using the small-molecule assay validation guidelines (US FDA 2018). Additionally, the impact of ADC stability (particularly at high concentration of ADC) on unconjugated payload during sample processing needs to be evaluated. For instance, since the expected unconjugated payload concentrations are comparatively lower, i.e., approximately less than 1% of molar concentration of ADC, even a 0.25% deconjugation of payload from ADC during storage can result in a 25% increase in unconjugated payload concentration.

3.5.2 *In Vivo DAR Characterization*

DAR characterization measures changes in the stoichiometric ratio between the Ab and payload in systemic circulation. Even with advancement in conjugation technologies, ADC deconjugation is still expected to occur in vivo resulting in changes in its DAR distribution over time relative to the dosed starting material. The ratio of conjugated payload and TAb analytes is defined as average DAR. In cases involving formation of an active metabolite due to the biotransformation of the payload, both the ratio of conjugated payload to TAb and conjugated-payload metabolite to TAb needs to be monitored. The change in average DAR of an ADC is an indicator of its in vivo stability.

Measurement of DAR in biological samples is challenging owing to the complexity of biological matrices and problems associated with desired high assay sensitivity. Methods involving affinity capture followed by intact LC-MS-based assay have been reported for in vivo DAR measurement (Xu et al. 2013; Xu et al. 2011; Su et al. 2016). This strategy, though efficient, suffers from low sensitivity due to the inherent low MS response of intact analysis. To enhance sensitivity, an alternative affinity LC-MS strategy has been reported, wherein the affinity captured ADC was cleaved to Fab and Fc fragments using immunoglobulin-degrading enzyme of *Streptococcus pyogenes* (IdeS) digestion prior to intact LC-MS analysis (Su et al. 2016). Similar methods using limited sample digestion have been reported for DAR characterization (Grafmuller et al. 2016; Davis et al. 2017; Rago et al. 2016). Either N-glycosidase F or IgGZERO® can be used for deglycosylation to reduce heterogeneity and simplify the mass spectra (Wei and Ramanathan 2019). But IdeS digestion is not suitable for DAR analysis of ADCs with lysine-based conjugation. While using the limited digestion workflow, it is necessary to ensure that the linker-payload will not be cleaved during sample preparation. A typical workflow of hybrid LC-MS/MS assays for DAR characterization of ADCs is illustrated in Fig. 3.4 (Adapted from sample preparation strategies for ADC bioanalysis described by Wei and Ramanathan 2019). Using intact MS analysis, relative ratios of individual ADC species are obtained based on their peak areas in the deconvoluted mass spectra. Average DAR is estimated with the calculation shown below.

$$\text{Average DAR} = \Sigma\left(\%\text{peak area} \times \text{Number of conjugated payload}\right)/100$$

A newer approach to calculating in vivo DAR is to separately measure conjugated payload and TAb using methodologies described earlier in this chapter. Higher sensitivity is achieved using this approach as it uses targeted peptide quantification instead of intact LC-MS analysis (Wang et al. 2016; Huang et al. 2020; Liu et al. 2015; Sanderson et al. 2016). However, this strategy is limited to ADCs with cleavable linkers.

3.6 Immunogenicity Assays for ADC

The development of an adaptive immune response (also referred to as immunogenicity response) against large-molecule drugs (aka biologics) is a well-known phenomenon and is typically monitored by measurement of ADA. As with other biologics, ADCs may also evoke immunogenicity response and hence require immunogenicity monitoring strategies driven by thorough immunogenicity risk assessment of the ADC molecule (US FDA 2018; US FDA 2019) (Fig. 3.5).

The immunogenicity risk factors unique to ADC include (Myler et al. 2019):

- ADA response(s) could develop against any functional component of ADC including mAb, linker, payload, or neo-epitopes exposed between these components.
- Hydrophobic nature of payloads can create aggregation-prone regions on ADC.
- ADC contain a hapten-like structure in which the otherwise non-immunogenic payload or peptide upon conjugation could render ADC immunogenic, a phenomenon known as hapten effect.
- ADC could also exhibit epitope spreading, a phenomenon where immune response is initially induced against one part of the molecule but over time diversifies to otherwise non-immunogenic domains/epitopes within the same protein.
- Multiple linker-payload structures on ADC could serve as repetitive antigenic structures that in combination with multivalent ADAs (especially pentameric immunoglobulin M (IgM)) could form large circulating immune complexes (CIC). Such CIC would be difficult to clear or be more likely to deposit in small vascular channels and interstitial spaces, resulting in immune-complex mediated diseases.

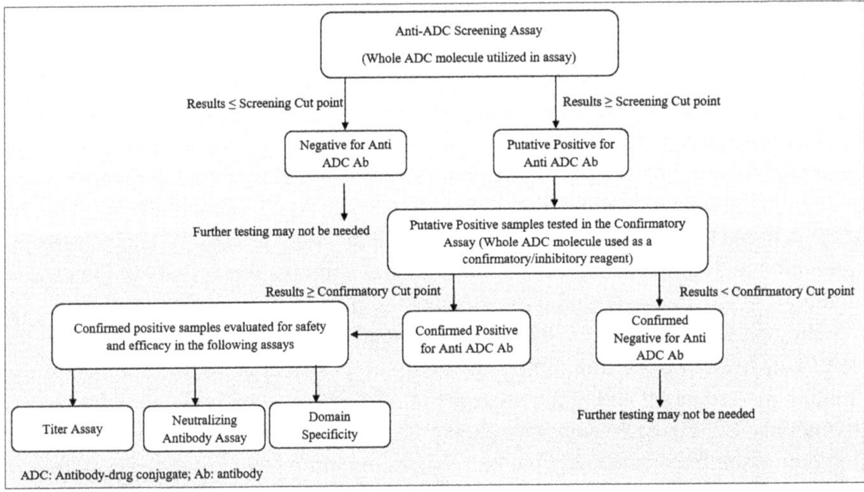

Fig. 3.5 Multi-tiered bioanalytical monitoring strategy for ADC immunogenicity assays

- Uptake of payload by non-target immune cells during large ADC-ADA immune complex clearance.

Although some of these risk factors are applicable for any biologics, they could potentially be enhanced by the multi-domain structure of ADC. Due to these unique risks associated with ADC and very limited clinical experience, ADCs are typically not classified under "low" immunogenicity risk category. Also, since the ADC-induced immune response would not cause neutralization of an essential endogenous counterpart, the ADCs are generally not perceived as "high" immunogenicity risk category either. Thus, ADCs are generally classified under the "medium" immunogenicity risk category (Myler et al. 2019; Hoofring et al. 2013). However, since every ADC is unique with distinct physicochemical attributes, linker-payload, and conjugation chemistry, each ADC may have unique associated risk factors that require careful consideration in the development of specific immunogenicity assay strategies for both nonclinical and clinical studies.

3.6.1 Immunogenicity Testing Strategy for ADC

An overall risk-based tiered immunogenicity assay strategy that is typical for biologics is also adopted for ADCs. The tiered strategy involves a stepwise approach where initially screening assays are used to detect ADA responses against any functional domain of ADCs including Ab, linker, payload, or neo-epitopes involving multiple ADC components. The immune response is then typically confirmed by competitive binding with the excess ADC. The immune response may be further characterized to (a) determine whether the response is primarily to the Ab or to the other components of ADC using various strategies, such as competitive binding with the unconjugated Ab and/or linker-payload, and (b) to determine the potential of the ADA to inhibit the ADC pharmacological activity.

Since the ADC immunogenicity assays must be capable of detecting ADAs against any portion of the ADC, including the neo-epitopes between different functional domains, it is recommended to use the entire ADC molecule for capture or detection in the screening and confirmatory methods. Positive controls (monoclonal Ab or polyclonal antibodies) directed against whole ADC and/or different domains (e.g., Ab, payload, or linker-payload) are used as assay suitability controls during validation and sample testing to confirm the immunological reactivity of the capture and detection of reagents against various types and subclasses of ADA.

The major challenges for ADC immunogenicity assays are the target tolerance and drug tolerance, i.e. the ability of the assay to detect ADA in the presence of soluble or shed target and in the presence of ADC therapeutic in the sample, respectively (USFDA 2018). Various sample pre-treatment approaches could be employed to remove soluble or shed target and/or therapeutic drug from test samples (Gorovits et al. 2014; Patton et al. 2005; Zoghbi et al. 2015). Another approach to reduce ADC therapeutic drug interference consists of selecting sample collection time points at the ADC trough levels during the clinical study.

3.6.2 Screening Assay for ADC

The screening assay aims to detect all subclasses of ADA and against all functional domains of ADC with appropriate sensitivity and tolerance to the ADC therapeutic in the sample. Assay formats such as direct assay and bridge assay formats that are used for screening assays for other biologics are well suited for ADC. The assay formats can be run on a variety of bioanalytical platforms with different readouts including ELISA, electrochemiluminescence (ECL), real-time biosensor-based methods, etc. Regardless of the bioanalytical platform, test samples with signal response equal to or above a statistically established threshold (also known as screening assay cut point) are determined as putative positive.

Homogeneous bridge assay format is the most widely used assay format for ADA assessments. In this assay format, samples are co-incubated with labeled-ADC as capture and detection reagents followed by immobilization of the ADA-labeled ADC complexes, typically using a streptavidin-coated solid phase and detection using solution phase ELISA or ECL readout.

3.6.3 Confirmatory Assay for ADC

The goal of the confirmatory assay is to refine the assessment of putative positive samples as defined by the screening assay. Confirmatory assay thus frequently uses the same assay format as the screening assay, with a competitive binding or competitive inhibition step. The excess unlabeled ADC is added to the test samples which screened putative positive. After an incubation period, the spiked and unspiked samples are analyzed with the screening assay. The percent signal reduction in the spiked sample compared to the signal in the unspiked sample is calculated. Samples with a percent signal reduction equal to or above the statistically calculated confirmatory assay cut point are considered confirmed positive.

3.6.4 *Titer Assay for ADC*

In titer assays, confirmed positive samples are serially diluted and tested in the screening assay. The reciprocal of the dilution at which the sample response is above the titer cut point is multiplied by the assay minimum required dilution (MRD) and reported as the titer value for the sample.

3.6.5 *Neutralizing Antibody Assay for ADC*

Neutralizing antibodies (NAbs) are a subpopulation of ADA that can potentially impact patient safety and directly mediate loss of drug efficacy by blocking the biological activity of ADC. Therefore, Nab detection is an important aspect of ADC immunogenicity assessment and is, by far, the most complicated aspect of the immunogenicity assessment.

Nabs against ADC can bind to any functional component of ADC. NAbs specific to the mAb or even the payload component (through steric hindrance) could block the binding of ADC to the target cells and/or inhibit cellular internalization due to formation of large immune complexes. In addition, NAbs against the payload could potentially inhibit its cytotoxic activity if they remain bound to the payload after the immune complex has been internalized and released from lysosome. Since the cytotoxicity or proliferation readout-based cellular assay incorporates critical steps of ADC MoA, the assessment of neutralizing capacity of ADA against ADCs is preferred using cell-based assays (Zhong et al. 2017).

An important caveat of such cell-based assays is that the sensitivity of cell lines to killing is a product of both target expression level and susceptibility to the payload and is not necessarily equivalent to those of target cells in vivo. In cases, where development of a cell-based assay with acceptable performance is impossible, competitive ligand-binding assays with improved sensitivity and drug tolerance could be considered appropriate, but timely discussion with regulators is strongly recommended.

3.6.6 *Domain Characterization Assay for ADC*

ADA specificity can be determined against the individual functional domains of ADC (mAb, linker, and payload). Two alternative approaches for ADA domain characterization against ADCs are - competition and direct detection.

In a competition approach, samples are incubated with excess unlabeled individual ADC functional domains and tested using the confirmatory assay format. Reduction of signal in presence of a specific ADC functional domain indicates specificity against that domain. In a direct detection method, specific labeled ADC domains are used for capture and/or detection of ADA against the respective ADC domain.

While both competition and direct detection methods can detect the most abundant ADAs, each has its own limitations (Myler et al. 2019). In the competition approach, it is possible that low abundant ADA against some ADC epitopes/domains may not be detected in the presence of high abundant ADAs to other epitopes/domains. In the direct detection approach, the molecular structures of the individual ADC domain reagents may differ slightly from the ADC, preventing evaluation of certain immune responses.

Both ADA domain characterization methods require generation of domain-specific reagents. While the unconjugated Ab is usually readily available, the high cytotoxic potency of payload and small size of the payload and linker complicate their use as unlabeled competition or labeled capture/detection reagents. Instead, the payload or linker are typically conjugated to a carrier protein. The use of an irrelevant Ab as a carrier is not recommended because of the expected high degree of homology to the constant regions of the Ab component of the ADC. Instead, an unrelated common protein, such as bovine serum albumin (BSA), is a preferable carrier for the payload/linker. However, if carrier protein conjugated linker/payload is used in the detection approach, then assay buffers should include excess carrier protein to inhibit the detection of non-specific antibodies against the carrier protein.

3.7 Future Perspective

While ADCs hold the potential for promising biotherapeutics that can broaden therapeutic index of chemotherapeutics. Overall, their clinical success has been plagued by the unexpected safety outcomes and/or payload resistance mechanisms. It has become increasingly clear that the clinical success of ADCs is dependent on the appropriate choice of target antigen, tumor type, linker-payload chemistry, and highly specific antibody.

The last decade has seen significant advancement in Ab engineering and bioconjugation methodologies resulting in next-generation ADC with better target binding affinities, stable linker chemistry, and payloads with higher potency. Newer bioanalytical techniques would thus need to be developed to address the evolving ADC landscape. For example, ADCs with novel and/or more potent payloads would require development of more sensitive assay with lower detection limits for quantification of payloads (conjugated and unconjugated). As the ADC protein scaffolds become more complex, such as bispecific, paratopic, or Probody™ antibodies, additional bioanalytical considerations would be required to monitor the integrity of the biotherapeutic molecule.

ADCs as a biotherapeutic class will continue to grow beyond cancer therapy in the future. Newer ADCs are being developed for non-oncological indications including autoimmune disorders, infectious diseases, cardiovascular diseases, and liver diseases (Liu et al. 2016, Shi et al. 2018). The ADC strategy, with its promising ability to achieve high specificity, high therapeutic index, and longer half-lives, provides an opportunity for drugs that have failed due to the lack of selectivity or undesired PK profiles.

In the future, improvements in ADC manufacturing and purification will enable development of standards that have specific DAR species that can be used for verification of bioanalytical assay performance to make more clinically relevant assays. Automation and artificial intelligence will enable more user-friendly instrumentation and sample preparation formats resulting in overall increase in throughput. Advances in the field of liquid chromatography and mass spectrometry, especially in the field of HRMS, for both intact protein analysis and proteomic-type experiments, will enable more comprehensive analysis of in vivo biotransformation and DAR distribution of ADCs.

Disclaimer Any opinions/forward-looking statements expressed in this chapter are those of the author(s) alone and may not reflect views held by their employers (Clinical Research Group, Thermo Fischer Scientific for Morse Faria; Inhibrx for Varun Ramani; EMD Serono, Inc. for Seema Kumar).

References

Alley SC, Benjamin DR, Jeffrey SC, Okeley NM, Meyer DL, Sanderson RJ, et al. Contribution of linker stability to the activities of anticancer immunoconjugates. Bioconjug Chem. 2008;19(3):759–65.

Heudi O, Barteau S, Picard F, Kretz O. Quantitative analysis of maytansinoid (DM1) in human serum by on-line solid phase extraction coupled with liquid chromatography tandem mass spectrometry-Method validation and its application to clinical samples. J. Pharm. Biomed. Anal. 2016;120:322–332.

Beck A, Goetsch L, Dumontet C, Corvaïa N. Strategies and challenges for the next generation of antibody–drug conjugates. Nat Rev Drug Discov. 2017;16(5):315–37.

Buckwalter M, Dowell JA, Korth-Bradley J, Gorovits B, Mayer PR. Pharmacokinetics of gemtuzumab ozogamicin as a single-agent treatment of pediatric patients with refractory or relapsed acute myeloid leukemia. J Clin Pharmacol. 2004;44(8):873–80.

Cahuzac H, Devel L. Analytical methods for the detection and quantification of ADCs in biological matrices. Pharmaceuticals. 2020;13(12):462.

Davis JA, Kagan M, Read J, Walles M, Hatsis P. Immunoprecipitation middle-up LC-MS for in vivo drug-to-antibody ratio determination for antibody-drug conjugates. Bioanalysis. 2017;9(20):1535–49.

DiJoseph JF, Armellino DC, Boghaert ER, Khandke K, Dougher MM, Sridharan L, Kunz A, Hamann PR, Gorovits B, Udata C, Moran JK. Antibody-targeted chemotherapy with CMC-544: a CD22-targeted immunoconjugate of calicheamicin for the treatment of B-lymphoid malignancies. Blood. 2004;103(5):1807–14.

Dong L, Li C, Locuson C, Chen S, Qian MG. A two-step immunocapture LC/MS/MS assay for plasma stability and payload migration assessment of cysteine-maleimide-based antibody drug conjugates. Anal Chem. 2018;90(10):5989–94.

Erickson HK, Widdison WC, Mayo MF, Whiteman K, Audette C, Wilhelm SD, Singh R. Tumor delivery and in vivo processing of disulfide-linked and thioether-linked antibody–maytansinoid conjugates. Bioconjug Chem. 2010;21(1):84–92.

Faria M, Halquist MS. Internal standards for absolute quantification of large molecules (proteins) from biological matrices by LC-MS/MS. calibration valid. Anal Methods A Sampl Curr Approaches. 2018:61. &ots=9wxIjGhKnY&sig=3dAz2naGTVLZ0Tp3MKr6_5oBh0U

Faria M, Peay M, Lam B, Ma E, Yuan M, Waldron M, Mylott WR, Liang M, Rosenbaum AI. Multiplex LC-MS/MS assays for clinical bioanalysis of MEDI4276, an antibody-drug conjugate of tubulysin analogue attached via cleavable linker to a biparatopic humanized antibody against HER-2. Antibodies. 2019a;8(1):11.

Faria M, Peay M, Lam B, Ma E, Yuan M, Waldron M, et al. Multiplex LC-MS/MS assays for clinical bioanalysis of MEDI4276, an antibody-drug conjugate of Tubulysin analogue attached via cleavable linker to a Biparatopic humanized antibody against HER-2. Antibodies. 2019b;8(1):11.

Furlong MT, Titsch C, Xu W, Jiang H, Jemal M, Zeng J. An exploratory universal LC-MS/MS assay for bioanalysis of hinge region-stabilized human IgG4 mAbs in clinical studies. Bioanalysis. 2014;6(13):1747–58.

Gorovits B, Alley SC, Bilic S, Booth B, Kaur S, Oldfield P, et al. Bioanalysis of antibody-drug conjugates: American Association of Pharmaceutical Scientists Antibody-Drug Conjugate Working Group position paper. Bioanalysis. 2013;5(9):997–1006.

Gorovits B, Wakshull E, Pillutla R, Xu Y, Manning MS, Goyal J. Recommendations for the characterization of immunogenicity response to multiple domain biotherapeutics. J Immunol Methods. 2014;(408):1–2.

Grafmuller L, Wei C, Ramanathan R, Barletta F, Steenwyk R, Tweed J. Unconjugated payload quantification and DAR characterization of antibody-drug conjugates using high-resolution MS. Bioanalysis. 2016;8(16):1663–78.

Hoofring SA, Lopez R, Hock MB, Kaliyaperumal A, Patel SK, Swanson SJ, et al. Immunogenicity testing strategy and bioanalytical assays for antibody-drug conjugates. Bioanalysis. 2013;5:9.

Huang Y, Del Nagro CJ, Balic K, Mylott WR, Ismaiel OA, Ma E, et al. Multifaceted bioanalytical methods for the comprehensive pharmacokinetic and catabolic assessment of MEDI3726, an anti-prostate-specific membrane antigen pyrrolobenzodiazepine antibody-drug conjugate. Anal Chem. 2020;92:16.

Huang Y, Mou S, Wang Y, Mu R, Liang M, Rosenbaum AI. Characterization of antibody-drug conjugate pharmacokinetics and in vivo biotransformation using quantitative intact LC-HRMS and surrogate analyte LC-MRM. Anal Chem. 2021;93(15):6135–44.

Jain N, Smith SW, Ghone S, Tomczuk B. Current ADC linker chemistry. Pharm Res. 2015;32(11):3526–40.

Jin W, Burton L, Moore I. LC-HRMS quantitation of intact antibody drug conjugate trastuzumab emtansine from rat plasma. Bioanalysis. 2018;10(11):851–62.

Junutula JR, Raab H, Clark S, Bhakta S, Leipold DD, Weir S, Chen Y, Simpson M, Tsai SP, Dennis MS, Lu Y. Site-specific conjugation of a cytotoxic drug to an antibody improves the therapeutic index. Nat Biotechnol. 2008;26(8):925–32.

Kamath AV, Iyer S. Challenges and advances in the assessment of the disposition of antibody-drug conjugates. Biopharm Drug Dispos. 2016;37(2):66–74.

Kaur S, Liu L, Cortes DF, Shao J, Jenkins R, Mylott WR, et al. Validation of a biotherapeutic immunoaffinity-LC-MS/MS assay in monkey serum: "plug-and-play" across seven molecules. Bioanalysis. 2016;8(15):1565–77.

Kaur S, Xu K, Saad OM, Dere RC, Carrasco-Triguero M. Bioanalytical assay strategies for the development of antibody-drug conjugate biotherapeutics. Bioanalysis. 2013;5(2):201–26.

Kumar S, King LE, Clark TH, Gorovits B. Antibody-drug conjugates nonclinical support: from early to late nonclinical bioanalysis using ligand-binding assays. Bioanalysis. 2015;7(13):1605–17.

Lee MV, Kaur S, Saad OM. Conjugation site influences antibody-conjugated drug PK assays: case studies for disulfide-linked, self-immolating next-generation antibody drug conjugates. Anal Chem. 2020;92(18):12168–75.

Lin K, Tibbitts J. Pharmacokinetic considerations for antibody drug conjugates. Pharm Res. 2012;29(9):2354–66.

Liu A, Kozhich A, Passmore D, Gu H, Wong R, Zambito F, Rangan VS, Myler H, Aubry AF, Arnold ME, Wang J. Quantitative bioanalysis of antibody-conjugated payload in monkey plasma using a hybrid immuno-capture LC–MS/MS approach: assay development, validation, and a case study. J Chromatogr B. 2015;1002:54–62.

Liu R, Wang RE, Wang F. Antibody-drug conjugates for non-oncological indications. Expert Opin Biol Ther. 2016;16(5):591–3.

Mou S, Huang Y, Rosenbaum AI. ADME considerations and bioanalytical strategies for pharmacokinetic assessments of antibody-drug conjugates. Antibodies. 2018;7(4):41.

Moucun Y, Ismaiel OA, Mylott WR. Hybrid ligand binding immunoaffinity-liquid chromatography/mass spectrometry for biotherapeutics and biomarker quantitation: how to develop a hybrid LBA-LC-MS/MS method for a protein? Rev Sep Sci. 2019;1(1):47–55.

Myler H, Gorovits B, Phillips K, Devanarayan V, Clements-Egan A, Gunn GR, et al. Report on the AAPS immunogenicity guidance forum. AAPS J. 2019;21:4.

Patton A, Mullenix MC, Swanson SJ, Koren E. An acid dissociation bridging ELISA for detection of antibodies directed against therapeutic proteins in the presence of antigen. J Immunol Methods. 2005;304:1–2.

Perez HL, Cardarelli PM, Deshpande S, Gangwar S, Schroeder GM, Vite GD, Borzilleri RM. Antibody–drug conjugates: current status and future directions. Drug Discov Today. 2014;19(7):869–81.

Polu KR, Lowman HB. Probody therapeutics for targeting antibodies to diseased tissue. Expert Opin Biol Ther. 2014;14(8):1049–53.

Rago B, Clark T, King L, Zhang J, Tumey LN, Li F, et al. Calculated conjugated payload from immunoassay and LC-MS intact protein analysis measurements of antibody-drug conjugate. Bioanalysis. 2016;8(21):2205–17.

Rago B, Tumey LN, Wei C, Barletta F, Clark T, Hansel S, et al. Quantitative conjugated payload measurement using enzymatic release of antibody-drug conjugate with cleavable linker. Bioconjug Chem. 2017;28(2):620–6.

Saad OM, Shen BQ, Xu K, Khojasteh SC, Girish S, Kaur S. Bioanalytical approaches for characterizing catabolism of antibody–drug conjugates. Bioanalysis. 2015;7(13):1583–604.

Sanderson RJ, Nicholas ND, Baker Lee C, Hengel SM, Lyon RP, Benjamin DR, et al. Antibody-conjugated drug assay for protease-cleavable antibody-drug conjugates. Bioanalysis. 2016;8(1):55–63.

Shen BQ, Bumbaca D, Saad O, Yue Q, Pastuskovas CV, Cyrus Khojasteh S, Tibbitts J, Kaur S, Wang B, Chu YW, LoRusso MP. Catabolic fate and pharmacokinetic characterization of trastuzumab emtansine (T-DM1): an emphasis on preclinical and clinical catabolism. Curr Drug Metab. 2012;13(7):901–10.

Shi C, Goldberg S, Lin T, Dudkin V, Widdison W, Harris L, et al. Bioanalytical workflow for novel scaffold protein-drug conjugates: quantitation of total Centyrin protein, conjugated Centyrin and free payload for Centyrin-drug conjugate in plasma and tissue samples using liquid chromatography-tandem mass spectrometry. Bioanalysis. 2018;10(20):1651–65.

Staudacher AH, Brown MP. Antibody drug conjugates and bystander killing: is antigen-dependent internalisation required? Br J Cancer. 2017;117(12):1736–42.

Stephan JP, Chan P, Lee C, Nelson C, Elliott JM, Bechtel C, et al. Anti-CD22-MCC-DM1 and MC-MMAF conjugates: impact of assay format on pharmacokinetic parameters determination. Bioconjug Chem. 2008;19(8):1673–83.

Stephan JP, Kozak KR, Wong WLT. Challenges in developing bioanalytical assays for characterization of antibody-drug conjugates. Bioanalysis. 2011;3(6):677–700.

Su D, Ng C, Khosraviani M, Yu SF, Cosino E, Kaur S, et al. Custom-designed affinity capture LC-MS F(ab′)2 assay for biotransformation assessment of site-specific antibody drug conjugates. Anal Chem. 2016;88(23):11340–6.

Tsuchikama KT, An Z. Antibody-drug conjugates: recent advances in conjugation and linker chemistries. Protein Cell. 2018;9(1):33–46.

US FDA Immunogenicity testing of therapeutic protein products—developing and validating assays for anti-drug antibodies, US FDA Guidance, 2019. www.fda.gov/regulatory-information/search-fda-guidance-documents/immunogenicity-testing-therapeutic-protein-products-developing-and-validating-assays-anti-drug.

USFDA Guidance for Industry Bioanalytical Method Validation Guidance for Industry Bioanalytical Method Validation," USFDA, May, pp. 1–44, 2018.

van den Broek I, Niessen WM, van Dongen WD. Bioanalytical LC–MS/MS of protein-based biopharmaceuticals. J Chromatogr B. 2013;929:161–79.

Wang J, Gu H, Liu A, Kozhich A, Rangan V, Myler H, et al. Antibody-drug conjugate bioanalysis using LB-LC-MS/MS hybrid assays: strategies, methodology and correlation to ligand-binding assays. Bioanalysis. 2016;8(13):1383–401.

Wei C, Ramanathan R. Sample preparation for LC-MS bioanalysis of antibody-drug conjugates. Sample Prep LC-MS Bioanal. 2019:335–50.

Wei C, Su D, Wang J, Jian W, Zhang D. LC–MS challenges in characterizing and quantifying monoclonal antibodies (mAb) and antibody-drug conjugates (ADC) in biological samples. Curr Pharmacol Rep. 2018;4(1):45–63.

Wei C, Zhang G, Clark T, Barletta F, Tumey LN, Rago B, Hansel S, Han X. Where did the linker-payload go? A quantitative investigation on the destination of the released linker-payload from an antibody-drug conjugate with a maleimide linker in plasma. Anal Chem. 2016;88(9):4979–86.

Wei WD, Sullivan M, Espinosa O, Yang L. A sensitive LC–MS/MS method for the determination of free maytansinoid DM4 concentrations—method development, validation, and application to the nonclinical studies of antitumor agent DM4 conjugated hu-anti-Cripto MAb B3F6 (B3F6-DM4) in rats and monkeys. Int J Mass Spectrom. 2012;312:53–60.

Xu K, Liu L, Dere R, Mai E, Erickson R, Hendricks A, et al. Characterization of the drug-to-antibody ratio distribution for antibody-drug conjugates in plasma/serum. Bioanalysis. 2013;5(9):1057–71.

Xu XK, Liu L, Saad OM, Baudys J, Williams L, Leipold D, Shen B, Raab H, Junutula JR, Kim A, Kaur S. Characterization of intact antibody–drug conjugates from plasma/serum in vivo by affinity capture capillary liquid chromatography–mass spectrometry. Anal Biochem. 2011;412(1):56–66.

Zhong ZD, Clements-Egan A, Gorovits B, Maia M, Sumner G, Theobald V, et al. Drug target interference in immunogenicity assays: recommendations and mitigation strategies. AAPS J. 2017;19:6.

Zoghbi J, Xu Y, Grabert R, Theobald V, Richards S. A breakthrough novel method to resolve the drug and target interference problem in immunogenicity assays. J Immunol Methods. 2015;426:62–9.

Chapter 4
An Introduction to Bioanalysis of Bispecific and Fusion Proteins

Kelly Covert, Hongmei Niu, and Sanjeev Bhardwaj

Abstract Bispecific antibodies and fusion proteins are part of the next generation of biotherapeutics that contain more than one functional domain. These biotherapeutic modalities are structurally complex and require unique approaches to bioanalysis. For their pharmacokinetic analysis, multiple assays may be required to characterize the disposition of the whole drug molecule as well as the individual functional domains. The immunogenicity analysis of these biotherapeutics also requires additional tiers of domain-specific characterization. The bioanalytical approaches for these modalities thus require additional considerations due to the complexity of their structure, additional domain-specific critical reagent generation, and increased assay complexities due to interferences observed by individual functional domains.

Keywords Bispecific antibody · Fusion protein · Pharmacokinetics · Immunogenicity · Neutralizing antibodies · Ligand binding assays · LCMS · Development · Validation

4.1 Introduction: Structural Overview, Comparisons, and History of bsAb and Fusion Proteins

Bispecific antibodies (bsAbs) and fusion proteins are structurally complex biotherapeutic agents that have the distinct advantage of being able to engage and act on multiple targets. Both are engineered structures, some utilizing existing antibody scaffolds that have been heavily modified, while others are genetically created through the expression of multiple genes. While the modalities are based on other familiar therapeutic formats, such as monoclonal antibodies, there are key

K. Covert (✉) · H. Niu
EMD Serono Research & Development Institute, Inc. (a business of Merck KGaA, Darmstadt, Germany), Billerica, MA, USA

S. Bhardwaj
Janssen Biotherapeutics, Spring House, PA, USA

S. Kumar (ed.), *An Introduction to Bioanalysis of Biopharmaceuticals*, AAPS Advances in the Pharmaceutical Sciences Series 57,
https://doi.org/10.1007/978-3-030-97193-9_4

differences in their structures, bioanalytical approach, and potential challenges faced during bioanalysis.

4.1.1 Bispecific Antibodies

Bispecific antibodies are a class of therapeutic antibodies recognized for their complex and variable structures, most notably the presence of two unique functional binding domains. While many monoclonal antibodies (mAbs) consist of two heavy chains and two light chains with identical antigen binding sites, bsAbs have two or more unique antigen binding sites, often resulting from individual combinations of the heavy and light chains or in some cases from genetic engineering to create two different arms of the antibody molecule. Regardless of synthesis approach, each of these domains can recognize and bind their own individual epitopes or antigens, giving the bispecific molecule the ability to recognize a novel combination of targets that cannot be replicated with just a monospecific monoclonal antibody treatment (Runcie et al. 2018).

Bispecific antibodies require some manipulation or genetic engineering to create the combination of the two distinct domains, and there are multiple subclasses dependent upon the mechanism of linking the domains and overall whole molecule structure. As bsAb engineering has become more advanced, the ability to custom design the domains to maximize preferred features has led to a wide variety of structural possibilities. There are currently over 80 different formats in use of bsAbs, but in general, they can largely be classified into two main subclasses: Fc containing and non-Fc containing. The structure of the molecule largely guides the bioanalytical approach that will be used for assay development for their pharmacokinetic (PK) and immunogenicity (primarily anti-drug antibody [ADA]) analysis.

Non-Fc-containing bsAbs are also referred to as fragment-based, as they do not contain a full IgG structure, unlike their Fc-containing bsAb counterparts. These molecules that lack the Fc region are often small molecules that rely on the antigen binding for all therapeutic efficacy. Due to the lack of Fc region in these fragment structures, there are noted disadvantages: shorter plasma half-life, lack of Fc-mediated activity, antibody aggregation, and observed stability issues (Labrijn et al. 2019). Potential workarounds to these drawbacks have been investigated, such as incorporating stabilizing reagents (i.e., human serum albumin) into the structure or creating a fusion to an Fc-body for an overall fusion protein structure.

Bispecific antibodies that do contain the Fc region are most similar to IgG-type antibodies, and inclusion of this familiar structure-type allows for use of known methods of purification and improved stability all associated with the common IgG structure. These molecules typically have two Fab arms and one Fc region. The Fc region may also be incorporated into the overall therapeutic approach, making the bsAb trifunctional due to the three distinct binding sites of the two independent Fab arms and the Fc region. The Fc region can play a critical role in increasing the antibody half-life (due in part to the increased size of the protein structure compared to those lacking the Fc region), as well as control effector functions related to antibody-dependent cell-mediated cytotoxicity (ADCC) and antibody-dependent cellular phagocytosis (ADCP). Of the Fc-containing bsAbs, there is further classification based on structure symmetry (asymmetric or symmetric) and modifications made to the Fc portion of the molecule to promote additional activity.

There are advantages and disadvantages to each of the individual bispecific structures possible, but these molecules are developed with specific goals based on the structural development. The result of binding multiple targets can have different results, such as immune cell recruiting, interference of receptor signaling, or forced-protein interactions. The MOA of the drug becomes important when designing the bioanalytical strategy, especially when considering what types of PK assays may be needed and how to correctly implement a functional cell-based neutralizing assay, if needed.

For bsAbs that rely on immune cell recruiting as the MOA, the bsAbs bind first to a target cell surface antigen on a tumor cell that is targeted by one arm of the molecule and then recruit and bind to T cells or natural killer cells with the other arm. Signaling interference may be achieved through the molecule interfering with ligand binding to multiple receptors, thus blocking activation of a targeted pathway. Lastly is the ability to force protein interactions through binding of two different proteins at once. Unlike blocking activation of pathways or interference, inducing protein-protein interactions with the presence of the bsAb can sometimes be the key to fixing a pathway. In cases where patients may be lacking a critical gene to produce a component in a signaling pathway, the bsAb can help fill in the missing piece and provide treatment for certain diseases (Kontermann and Brinkmann 2015).

The modern-day push for bispecific and fusion protein development has been partially fueled by the previous challenges encountered when presenting combination therapies (i.e., multiple mAbs with different targets that are co-administered in one treatment). For a combination therapy to receive approval, it must include detailed justification for the use of the combination of multiple mAbs, including clinical and nonclinical data which support that the combination benefits exceed the individual mAb therapeutic efficacy. There is heightened focus on the potential increase to toxicity profiles and awareness of the effects of both (or multiple) therapies as opposed to the risk with only monotherapy treatment. BsAbs have the structural advantage of targeting, activating, or impacting multiple pathways without the increased risk or research required to support combination therapies. Additionally, there is a prevailing theory referred to as the "avidity hypothesis" which surmises there is a correlation in the increase in efficacy of BsAbs directly related to the increase in receptor-types bound. The ability of a bsAb to bind two different receptor-types at once increases the biological and mechanical impact of the drug, improving efficacy over two separate monotherapies binding independently (Trivedi et al. 2017).

The first approved bispecific antibody in the USA was Amgen's blinatumomab (Blincyto®), which has been utilized in the treatment of patients with B-cell acute lymphoblastic leukemia (B-ALL) and B-cell non-Hodgkin's lymphoma. It is a dual-target CD19- and CD3-bispecific molecule that utilizes the immune cell recruiting pathway (Mullard 2015). The mechanism of action for the molecule is to target CD19 expressed on the surface of malignant cells while simultaneously targeting and binding the CD3 receptor on T-cells. When the CD3 receptor is engaged and the T-cell is activated, it can enact cytotoxic functions on the cancerous cell in attempts to eradicate tumor formation. The second approved bsAb is emicizumab (Hemlibra®), which has been utilized in the treatment of patients with hemophilia A (Knight and Callaghan 2018). The molecule binds to both coagulation factor IX and coagulation factor X, replicating the action of coagulation factor VIII, which is not present in patients with hemophilia A. Utilizing the protein-protein interaction method of action, it is able to activate factor X through the forced proximity of factor IX and results in potentially reduced number of joint bleeds and bleeding rate for the patient.

Since the approval of blinatumomab and emicizumab, multiple bsAbs have been started in clinical studies to investigate the potential pathways for fighting and eradicating many different types of diseases. The bsAb market was expected to exceed $10 billion in the USA in 2020, with over 300 new bsAb drugs in the current market pipeline. While final estimates for 2021 are still pending, understanding the background and structure of the molecule is critical as it guides the understanding of the impact on the ADME properties and approaches to be used for bioanalysis.

4.1.2 *Fusion Proteins*

According to the American Medical Association, "A fusion protein is defined as a multifunctional protein derived from a single nucleotide sequence that may contain two or more genes or portions of genes with or without amino acid linker sequences. The genes should originally code for separate proteins, both with pharmacological action (e.g., action and targeting)." In other words, "A fusion protein (or chimeric protein) is produced by the expression of two or more linked genes or gene fragments that otherwise encode different proteins or protein segments" (CAS & National Science Library). Since the first fusion protein drug etanercept (Enbrel®) was approved in 1998 by US FDA (US Food and Drug Administration), fusion protein biotherapeutics have become one of the top four most profitable classes of biotherapeutics, and the total global sales revenue has been increasing at a rate of 5.5–6% annually (CAS & National Science Library).

Similar to bsAbs, an advantage of fusion proteins is their ability to fuse two or more protein domains, which provides a wide opportunity to generate novel combinations of functions to improve PK and PD properties. Additionally, this increases the breadth of target diversity driven by therapeutic purpose when compared to the regular recombinant proteins with a single domain.

In general, fusion proteins are mainly divided into two categories based on the binding properties of the linked component: half-life extension based and non-half-life extending activity based. Proteins in the half-life extension-based category utilize the linked components such as Fc-fusion, albumin fusion, transferrin fusion, carboxy-terminal peptide (CTP) fusions, and XTEN fusion, as well as others such as elastin-like peptide (ELP). Table 4.1 describes the mechanism of action for each type of half-life extension-based fusion protein. Among the approved drugs, etanercept (Enbrel®) and aflibercept (Eylea®) are two examples of the Fc-based fusion proteins; albutrepenonacog alfa (Idelvion®) which was approved by the FDA in 2016 for treating hemophilia B belongs to albumin-based fusion protein (US Food and Drug Administration), and corifollitropin alfa (Elonva®) received EU marketing approval in 2010 and was produced by fusing CTP with follicle-stimulating hormone (European Medicines Agency).

Fusion proteins in the second category of non-half-life-extending activity may contain two or more non-half-life-extending active components derived from different proteins or peptides. The active components usually have multiple target-binding specificities: binding to the soluble target or binding to the cell membrane target. Of them, the active components which bind to cell membrane could also serve as a guide to allow the fusion proteins to be more specifically delivered to targeted cells, thereby increasing the drug specificity and minimizing side effects. For example,

Table 4.1 Description of mechanism of action

Half-life extension-based fusion protein	Description	Mechanism of action	Citation
Fc-fusion	Fc portion of human IgG	Binding of Fc of human IgG to FcRn to affect the recycling of Fc fused protein by epithelial cells	Roopenian DC (2007)
Albumin fusion	66.5 KDa of human serum albumin (HSA)	Retarding the drug's filtration via kidney: Upper limit of size for glomerular filtration of proteins by the kidney; being the strong anion binding to FcRn and being recycled similar to IgG	Paolo Caliceti (2003), Roopenian DC (2007)
Transferrin fusion	Highly abundant serum glycoprotein 3–4 mg/mL in serum About 80KDa in size	A clathrin-dependent transferrin receptor-mediated mechanism which recycles transferrin receptor-bound transferrin back into the circulation	Xiaoying Chen (2011)
Carboxy-terminal peptide (CTP) fusions	Carboxy-terminal peptide	Impairing renal clearance via the negative charge and heavy sialylation	F A Fares (1992)
XTEN fusion	An amino acid repeating polymer	Some level of shielding effect, similar to PEG	Volker Schellenberger (2009)
Others	e.g., ELPylation (elastin-like peptide)	Giving fusion proteins a larger hydrodynamic radius, and thus they are not eliminated by the kidney	Conrad (2011)

denileukin diftitox is an FDA-approved fusion protein drug comprised of modified diphtheria toxin and interleukin 2 (IL-2), in which the diphtheria toxin functions as a toxophore and IL-2 acts as a targeting moiety to deliver the toxin to CD25-bearing cancer cells due to CD25's high affinity for IL-2 (CAS & National Science Library). Currently, development of bi–/multifunctional fusion proteins has become a major focus of ongoing therapeutic fusion protein research and development. As was discussed with bsAbs, the structure and MOA of fusion proteins are important characteristics to understand of the therapeutic in order to appropriately design and carry out successful bioanalysis.

With novel therapeutic modalities evolving, there is some overlap between fusion proteins and bispecific antibodies. Recently, around 40 percent of therapeutic fusion proteins have been reported to contain at least one antibody-related component (target-binding moiety) (CAS & National Science Library). In this regard, bispecific/multi-specific antibodies can be viewed as an extensive application of fusion proteins to enhance the function of traditional antibodies for multi-targeted therapies.

4.1.3 ADME

The absorption, distribution, metabolism, and excretion (ADME) properties of a drug are highly dependent on the drug structure and composition. Given the complexity engineered into the bsAbs and fusion proteins, it is expected that the ADME characteristics would be affected when compared to similar monospecific monoclonal counterparts. As there are multiple active components of these proteins, it is important to measure not only the pharmacologically relevant properties of the whole drug but also the individual functional domains as well.

There are many factors which can influence the ADME of a molecule, such as molecular mass, molecular charge, and protein modification. In the case of bsAbs, the overall molecular mass and size may not be significantly different from that of a monospecific if the structure has not significantly changed, as seen in the cases of the IgG-like bsAbs; however, extensive modification to the standard structure with multivariable Fab segments may lead to an increase in overall size. Conversely, the non-IgG bsAbs, especially small molecules or fragment-based structures, will have notably different ADME due to the significant changes in overall size. These smaller-sized bsAbs notoriously have shorter half-lives which impact the PK properties of the molecule. To achieve desired PK properties, these smaller-sized bsAbs may require modification to increase half-life, such as labeling with PEG or conjugating to albumin to increase size and structure stability. Modifications made to increase half-life and prolong drug exposure can severely impact the ADME of the molecule, and once these modifications are made, the bsAb must be reevaluated for impact to the ADME characteristics, specifically absorption and distribution. Multiple studies have been conducted on molecules modified with PEG, and they have shown that drug distribution is impacted, but not in a consistent, trending matter, so assumptions or extrapolations cannot be made to the individual ADME impact to each molecule after the modification without performing new investigations (Tibbitts et al. 2016).

Target binding can also impact the ADME of the molecule, and with the ability to bind two or more targets, bsAbs may see significant changes to each of these characteristics with each target binding. With target binding that also engages modes of action specific to bispecific molecules—such as immune cell activation and protein-protein interaction—the intended therapeutic pathway may also be what negatively impacts the ADME (Tibbitts et al. 2016).

One recent study (Datta-Mannan et al. 2016) investigated the differences in ADME characteristics of bsAbs compared to similar monospecific mAbs. The study evaluated the biodistribution and pharmacokinetic information of two IgG-like bsAbs in vivo (cynomolgus monkeys) and found both suffered from a rapid clearance and excretion and a shortened half-life, when compared to similar mAbs. The bsAbs showed up to tenfold increased faster elimination and up to fivefold increased half-life when compared to the parent mAb. While usual suspicions of the reduced properties may be target binding, Fc receptor binding, or overall molecular properties, it was determined these were not the cause of the impacted ADME characteristics. Upon ruling out these potential causes, new hypotheses were formed with refocused attention on the binding of the bsAb with liver sinusoidal endothelial cells (LSECs), instead of macrophages, and potential conformational changes of the molecule in vivo. The study highlighted the complexity of characterizing ADME properties of bsAbs and reiterated that they cannot be assumed to be the same as their mAb counterparts.

As mentioned in the fusion proteins structure section, there are two classes of fusion proteins: those that extend half-life and those that do not. This classification directly impacts ADME characteristics and the measurable PK properties. Knowing that both bsAb and fusion proteins will differ from other modalities in ADME characteristics, specific bioanalytical methods must be developed that are tailored to their structure and classification types.

4.2 Bioanalytical Method Development and Scope

4.2.1 Pharmacokinetic (PK) Approach

4.2.1.1 Bioanalytical Strategy

The pharmacokinetic and pharmacodynamic (PK/PD) characterization is an integral part of the drug development process. The bioanalytical strategy for bsAb and fusion proteins varies depending on the stage of drug development (preclinical versus clinical and what phase of clinical analysis), the availability of critical reagents and the PK question that is being asked at the given stage of the program. In general, the ligand binding assay (LBA) format has been the gold standard for quantification of biotherapeutics and characterization of immune response to biotherapeutics (Swann and Shapiro 2011; Kaur et al., 2013; Stevenson et al., 2014). Therefore, this section will discuss the application of LBA platform for bsAb and fusion protein PK characterization.

4.2.1.2 Critical Reagent Generation

Anti-idiotypic mAbs with neutralizing or non-neutralizing activities play a critical role to support LBA-based specific PK assay method to quantify intact target-free or total (target-free, partially bound and fully bound) drug. In terms of the generation of anti-idiotypic antibody specific to fusion proteins, a few things need to be taken into consideration such as the timing, technologies, immunogens, and antibody clone screening approaches. Generally speaking, the lead time for anti-idiotypic mAb generation ranges from 3 to 9 months depending on the selected technologies and the complexity of biotherapeutic structure; therefore, the bioanalytical strategy needs to be built to accommodate reagent generation time to the overall timeline to save the cost and frustration along the line. The main technology platforms for anti-idiotypic mAb generation involve two types: hybridoma and phage display. Regardless of the technology platforms, the utilized antigen and clone screening methods are comparable to some extent. As fusion proteins contain multi-specific domains, the immunogen used for immunization or phage display can be either intact molecule or each domain of fusion proteins. In this regard, an appropriate screening approach is very critical to enable identification of antibody clones which bind to each domain of fusion proteins in either domain fragment or intact drug form and can compete or non-compete target binding to the drug. In early assay development for nonclinical bioanalysis, generation of custom antibody reagents may not be feasible due to timing and budget constraints. Use of commercially available target protein and generic detection reagents can be utilized to create an acceptable assay format in early-stage development. As the potential therapeutic progresses to late-stage nonclinical testing or clinical testing, the generation of the specific reagents becomes more useful as assay optimization characteristics, such as sensitivity and interference testing, can benefit from these specific, non-commercial reagents to enhance the overall assay use.

4.2.1.3 Platforms

Much of the bioanalytical strategy discussed for fusion and bispecific proteins has been reliant upon plate-based applications. The most common approach for PK analysis for these modalities is through plate-based analysis performed on microplate readers that measure absorbance and optical density, or the Meso Scale Discovery (MSD) platform. LCMS applications to bispecific and fusion proteins are very limited in literature, likely due to the complex nature of these drug candidates (Cao et al. 2018; Iwamoto et al. 2018; Iwamoto and Shimada 2019; Kaur et al. 2016; Ma et al. 2019). The drug characterization, purity, and stability are the main focus of most LCMS workflows. Establishing the purity of the drug product is the biggest challenge facing the current bsAb and fusion proteins-based drugs. LCMS basics and instrumentation have been discussed in detail in previous chapters and so will not be repeated in this chapter. Immunoaffinity capture using an anti-idiotypic antibody followed by LCMS of the captured drug would be most appropriate for these modalities. The sample preparation and separation/resolution are critical to these workflows, and thus should be vetted thoroughly. Differential/ion mobility separations and capillary electrophoresis are a some of the front-end instrument modifications to explore the molecular structure of drugs and thus help with the full characterization of the drug candidate (Iwamoto and Shimada 2019).

The intact bsAb assay to measure the active form of the drug would require either the target or anti-idiotypic antibodies for binding to both arms of the bsAb (one arm for the capture Ab and the second one for binding to the detection Ab). LCMS could be applied to quantify the intact drug after immunoaffinity pulldown with an anti-id, but assay sensitivity could be hampered by the size of the drug and matrix effects.

The functional domain assay helps determine the active domain and informs of the loss of function due to ADA formation or biotransformation (Ma et al. 2019; Wang et al. 2016). The assay design for the measurement of the free domain would be based on the targeted domain and thus would have either the relevant therapeutic target or blocking anti-id. The immunoaffinity capture of bsAb using anti-ids for either of the arms followed by LCMS would help with understanding of the in vivo or in vitro biotransformation if suspected for the drug molecule.

Total assay would enable measurement of all forms of the bsAb (free, partially, and fully bound). The choice of the reagents would have a large impact on success of this assay design. Usually, a couple of antibodies specific to different regions/epitopes of the Fc would be used as capture and detection antibodies for an LBA assay design. The assay has limited utility and thus could be used in nonclinical stages. LCMS could be used for the measurement of the total bsAb in nonclinical and clinical stages of development. The success of the assay would depend on an anti-Fc antibody (for nonclinical) and anti-id targeting the Fab region (clinical) to pull down all forms of bsAb for measurement of total drug in subsequent steps of the LCMS workflows (Ma et al. 2019; Li et al. 2012).

The mis-paired product formation during the manufacture of the bispecific antibodies has to be fully characterized in order to minimize side products and optimize the manufacturing process (Cao et al. 2018). There are several technical platforms that have been utilized to address this issue. LCMS has been the platform of choice because of its ease of use and high-throughput analysis. The complexity in the structure of the drug candidates warrants thorough structural characterization to help with suspect biotransformation during in vivo preclinical assessments. The biotransformation process may break down the drug molecule into different subunits via clipping, deamidation, or oxidation, for example, which can result in varying concentrations of the whole drug molecule compared to the processed subunits. The downstream development of drugs can benefit by determination of the ratios of main drug vs the mixed arm bispecific subunits that may result from the processed portions of the fusion protein (Wang et al. 2018; Haraya et al. 2019; Jeremy Woods et al. 2013). The purity assessment of the drug candidate is a critical component of the successful therapeutic drug development. LCMS have been widely used as the technology of the choice for such applications. One should be able to discern and quantify the misaligned dimers as well as any truncations associated with the final product. Intact mass analysis coupled to the peptide mapping workflows was able to fully characterize the impurities due to undesired homo-/heterodimers as well as the truncated final products. These observations should help optimize the manufacturing of the main heterodimer or sc-Fv conjugates, thus minimizing the impact of the impurities (Jeremy Woods et al. 2013).

A couple of case studies on bispecific antibody PK challenges (Ma et al. 2019) during nonclinical phase emphasize the importance of orthogonal techniques such as LCMS for development and troubleshooting of PK/ADA assays. The poor efficacy of the drug in the PK assay results for a bispecific was attributed to either ADA impact (increased elimination) or biotransformation (truncations, glycosylation pattern, charge variants) affecting the drug concentration measurement in the total PK assay for the drug. The LCMS data helped eliminate biotransformation products as the cause for the anomaly. Higher levels of the ADA vs parent were observed and thus implicated for the anomaly in the total PK assay results.

PK analysis of the F(ab')2 biotransformation into 2 active F(ab) monomers in vivo is another example where LCMS would help identify and characterize the in vivo biotransformation and thus help guide the PK assay development and bioanalysis. A post-translational modification (PTM) in one arm of a bispecific antibody was observed during the in vivo biophysical characterization of the drug molecule. The point mutation at the active site resulted in the loss of target binding at that arm and thus may cause accumulation of the inactive drug that may lead to toxicity by repeat dosing. The in vivo characterization of the PTM using LCMS would be needed for further development of the bispecific. The active drug and total assay (free and bound) would help determine the amount of the inactive drug with point mutation. A target binding ligand binding assay would be most suitable for the "active" drug quantification as binding to a recombinant target or target surrogate is accepted as a valid surrogate for drug activity. The total drug quantification assay based on peptide quantification would be suitable for the "total" drug measurement. The findings in turn would enable future dosing in nonclinical/clinical stages of the drug development (Kellie et al. 2020, Giorgetti et al. 2020).

4.2.1.4 Clinical Vs. Nonclinical

Generally speaking, a generic PK assay is a common approach during preclinical stage of drug development which employs reagents specific to human proteins that don't cross-react with non-human species. This can be a fast and cost-effective approach to measure total protein which is suitable for supporting overall toxicological assessment (Ma et al. 2019), especially for half-life extending fusion proteins utilizing the linked components such as Fc-fusion. In that regard, commercial reagents targeting the known components such as using anti-human Fc antibody as a capture and detection reagent will be sufficient to detect the total Fc fusion protein drug. Additionally, a target for one binding domain fusion protein can serve as capture reagent with anti-protein framework (such as anti-Fc antibody) can serve as detection reagent free drug needs to be measured during the nonclinical stage.

Adrug-specific detection method is more likely necessary during clinical drug development, as this often demands detection of intact fusion proteins and, in certain scenarios, prevents interference due to the cross-reactivity of generic assay reagents with endogenous molecules. For example, to quantitate Fc-fusion protein drug, the assay sensitivity and specificity could be compromised if anti-human Fc antibody is used as a capture and/or detection reagent due to the presence of a large amount of human IgG in human blood. In principle, a specific LBA PK assay heavily relies on specific reagents for each domain of fusion proteins such as targets or anti-idiotype antibodies.

A few formats of drug exist in human matrix including free drug or drug-target complexes if the target is a soluble ligand or shed receptor. Theoretically, LBA format can be designed to measure all forms of drug when using proper assay reagents. In absence of or with low amounts of soluble ligand/shed receptor, total and free drug species are often equivalent, and their detection is less sensitive to assay formats or reagent choices (Fischer et al., 2012). In this case, an assay format employs a target or anti-idiotype antibody (non-neutralizing or neutralizing) specific to each domain of fusion proteins to quantify the total or free intact drug. In contrast, in the presence of a significant amount of soluble target, the quantification of free intact drug may need to utilize target or neutralizing anti-idiotypic antibody for each binding domain of fusion proteins, while quantification of total intact drug has to utilize a combination of non-neutralizing anti-idiotypic antibodies against each domain. In general, anti-idiotypic antibodies are more likely preferred rather than the targets in consideration of the relatively more stable feature of antibody structure.

Although there is still an ongoing debate as to what form of the drug is more relevant to measure, measurement of free intact drug is more relevant to establish PK-PD relationships in clinical phase. Therefore, reagents specific to each binding domains involving target and/or neutralizing or non-neutralizing anti-idiotypic antibody determined by the interference of endogenous soluble target as discussed above are utilized as the capture or detection reagents to measure free molecules that contain both functional domains.

The drug's MOA is also imperative to determine whether an intact drug or only one functional domain of fusion proteins needs to be measured based on either one or two domains engage in therapeutic activities. To some extent, non-half-life-extending fusion proteins more likely fall into in the first category to measure the intact drug where measuring the intact active fusion proteins is the best strategy to fully characterize the molecule, while for half-life-extending fusion proteins, using one drug-specific reagent to measure a single domain of fusion protein would be a reasonable strategy (e.g., Fc fusion proteins). In this case, therapeutic activity is dependent upon engagement of one functional domain, and the function of the second domain is to enhance the therapeutic activities leading to measurement of intact drug less important. In this case, the assay format for quantifying one function domain of fusion protein employs the therapeutic target or neutralizing anti-idiotype antibody specific to one domain of fusion proteins coupled with anti-framework antibody. Overall, measurement of intact fusion protein drug using reagents specific to each functional component is strongly recommended if specific reagents have been generated and are available for assay development, especially during clinical stage (Manikwar et al. (2020); Sawyer et al. (2020); Voronin et al. (2014)).

In general, the principle of bioanalytical strategies for fusion protein can be applied to bispecific therapeutics, which can be viewed as an extensive application of fusion protein containing antibody- or scaffold-based proteins. Like fusion proteins, a generic PK assay may be utilized during preclinical phase for bispecific development, while in the clinical phase, more drug-specific reagents (e.g., targets and anti-idiotype antibodies) are more likely necessary, and measurement of free drug is more relevant to PK-PD relationship. Likewise, measuring the intact active bispecific drug or one arm of bispecific is driven by the drug's MoA where the therapeutic activities depend on engagement of one or both targets. Figure 4.1 is a good summary for different assay formats and reagents to quantify bispecific therapeutics (Ma et al. 2019) (Fig. 4.2).

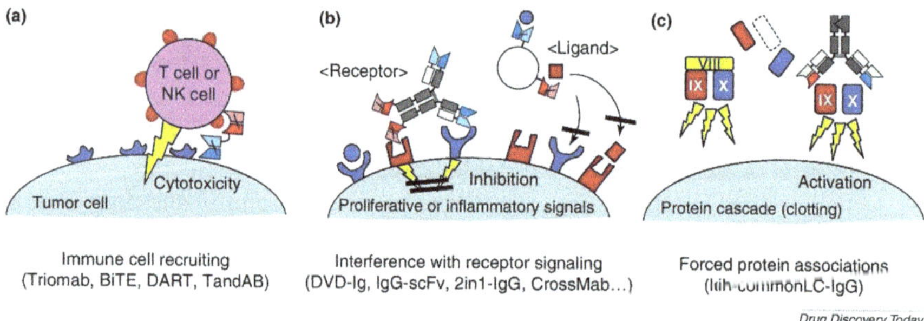

Fig. 4.1 Mode of action of therapeutic bispecific antibodies. From Kontermann and Brinkmann, "(**a**) Recruiting of T cells or natural killer (NK) cells to tumors is achieved by entities that bind to tumor cell surface antigens as well as to immune cells. Examples are TrioMabs (catumaxomab), BiTEs (blinatumomab), DARTs, and TandAbs. (**b**) Interference with receptor signaling is achieved by binding cell surface receptors or to their cognate ligands. BsAbs in various formats have been developed for this mode of action, such as DVD-Igs, DAFs, 2-in1-IgG, Tv-IgGs, and CrossMabs. (**c**) One exciting 'unusual' application of bsAbs is antibody-mediated forced assembly of the coagulation Xase complex. A heterodimeric common light-chain IgG connects FXIa and FX and thereby overcomes FVIII deficiency. Abbreviations: BiTE, bispecific T cell engager; DAF, dual-action Fab; DART, dual-affinity retargeting; DNL, dock-and-lock; DVD-Ig, dual-variable domain immunoglobulins; FX, factor X; HSA, human serum albumin; Ig, immunoglobulin; kih, knobs into holes; Tv, tetravalent"

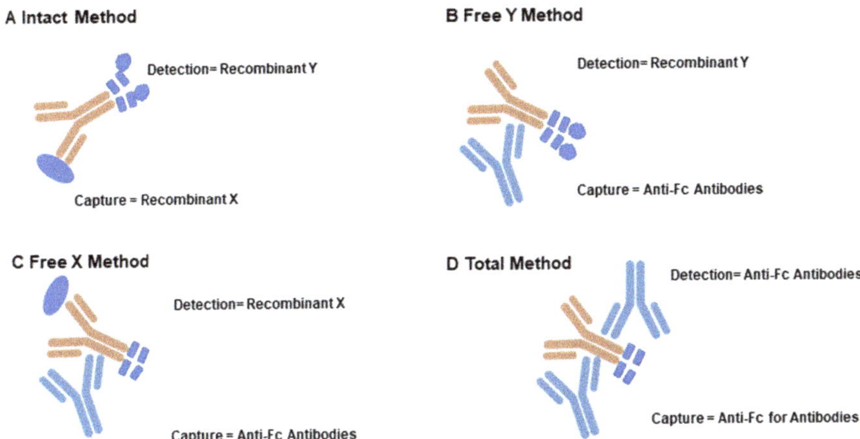

Fig. 4.2 Different assay formats and reagents for quantification of bispecific molecules. (**a**) Intact method: The intact bispecific molecules are measured by employing both target X protein and target Y protein as assay reagents. (**b**) Free Y method: The concentration of functional domain Y is measured by employing anti-Fc antibodies and target Y protein as assay reagents. (**c**) Free X method: The concentration of functional domain X is measured by employing anti-Fc antibodies and target X protein as assay reagents (**d**) Total method: All soluble forms of a bispecific molecule are measured by employing two anti-Fc antibodies as assay reagents

4.2.1.5 Challenges in PK Bioanalysis

A novel construct of fusion proteins and bispecific antibodies may pose increased immunogenicity risk which leads to drug-ADA complexes in a bio-matrix sample. For PK/PD evaluations, the primary interest is the non-neutralizing ADA-bound drug because it is the bioactive form, and the PK assay can be designed with employment of neutralizing anti-idiotypic antibody or therapeutic target as capture reagent (Wang et al., 2014).

As with other large molecules, PK method validation for fusion protein and bispecific molecules need to meet regulatory expectation and demonstrate that the following assay parameters meet assay acceptance criteria: precision and accuracy, linearity of dilution, selectivity, interference and specificity, sensitivity, and stability testing to ensure the assay suitable for the intended use to accurately and precisely measure drug concentration in human matrix. However, due to the nature of fusion protein and bispecific molecules which contain multiple functional domains, specificity and assay interference evaluation for an intact assay should take both domains into consideration (Ma et al. 2019). While this will extend the scope of validation activities, it provides a more complete picture of the PK performance of the entire molecule.

4.2.2 *Immunogenicity and Anti-Drug Antibody (ADA) Approach*

4.2.2.1 Bioanalytical Strategy

Given the complex structures associated with both bsAbs and fusion proteins, the rate of immunogenicity must be closely studied based on all aspects of the molecule. The unique structures guide all facets of the bioanalytical strategy from reagent generation, preferred platforms, and challenges that must be addressed. Understanding of the drug structure and sequence is critical in predicting the immunogenicity prior to initiation of testing and helps in identifying a molecule as "high risk" if it is expected to have a higher rate of immunogenicity compared to a less complex modality. Taking structure complexity into consideration, asymmetric bsAbs are believed to have the lowest rate of immunogenicity as their structure most closely resembles naturally occurring antibodies and modifications are less severe. Bispecific antibodies and fusion proteins that have complex, non-native antibody domains and linker sequences not seen in natural occurring antibodies have been shown to have higher rates of immunogenicity as immune systems do not easily recognize these structures (Labrijn et al. 2019). This would include non-Fc-containing bispecifics (fragment based), Fc-containing bispecifics with appendages (IgG-like, but with additional fragment attached to the IgG structure), and fusion proteins that do not share homologous sequences with endogenous counterparts. Another additional risk factor for increased immunogenicity are any novel epitopes that bsAbs and fusion proteins may target. With the various mechanisms of action available to these therapeutics, there is a heightened possibility of evoking an unknown immunogenic response when a new target is engaged.

In silico or simulation-based immunogenicity analysis may be performed based on the protein's sequence to determine if the molecule is at an increased risk for eliciting immunogenic responses and determining the correct strategy for identifying neutralizing antibodies. In general, the closer the molecule is to a fully humanized antibody structure and when components represent endogenous counterparts closely, there is a lower risk for immunogenicity. Based on the determined risk of immunogenic responses, the approach to measuring immunogenicity can be customized: if there is low risk of immunogenicity, then a dose escalation plan may proceed without reviewing immunogenicity data in the interim; higher-risk molecules may need to await the immunogenicity results of initial dosing before proceeding (Kernstock et al. 2020). As for development and implementation of nAb testing, if early phase testing with PD markers indicate sufficient monitoring of nAbs, then a competitive ligand binding nAb assay may not be necessary, but if response from regulatory agencies indicate otherwise, then there should be preparation for a nAb assay to be in place to support phase 3 studies at minimum.

The general bioanalytical approach to measuring immunogenicity starts similar to that of mAbs, with an initial tier for screening for the presence of anti-drug antibodies. Typically, this is performed using a plate-based format on ELISA or ECL platforms, utilizing critical reagents specifically labeled to the drug product for ADA detection. Note that in some instances, the standard plate-based format may not be appropriate based on target interference or failure of the molecule to easily follow the bridging format. Additional platforms or assay procedures may be followed and are described in detail in the future sections. Regardless of assay setup, screening for ADA presence to the whole molecule must be performed first.

The second tier of analysis is to confirm that the ADA response detected in the screening assay is specific to the drug molecule. This tier is usually performed by treating the samples with excess drug and measuring the inhibition of signal compared to samples that have not been treated with excess drug. In this approach, if significant reduction or inhibition of signal is observed in the presence of drug, then it is likely that the ADA response detected is binding to the drug in the confirmatory tier and is a specific antibody to the drug treatment. Appropriate levels of reduction or signal inhibition may be determined based off negative sample inhibition and establishing an appropriate cut point taking background changes into consideration. The confirmatory tier is performed with the whole molecule (fusion or bispecific) binding to the present ADAs. In further tiers for characterization of antibody response, fragments of the whole molecule may be used for additional information gathering.

Once an antibody response is confirmed specific to the drug, a third tier of bioanalytical testing may be performed utilizing a titer format. This is typically performed in the same screening assay format as tier one, but the samples are analyzed at multiple dilutions (titers) to determine how strong the antibody response is in the samples. Usually none of these tiers are considered quantitative and instead are only qualitative in measurement; results are positive or negative for antibody screening, confirmed or not confirmed for drug inhibition, and then there may be a semi-quantitative titer response, but it is generally not reported in relation to a known concentration of ADA.

As the overall approach is similar to that of other modalities, the parameters for assay validation also fall in line: establish appropriate cut points based on specified false-positive rates for each tier; determine sensitivity, drug tolerance, precision, tolerance, and interference of existing matrix factors that may be present in a sample; evaluate hook effect; and determine stability for freeze thaw cycling, benchtop, and any other sample handling conditions that may be necessary. While many of the official guidance do not specifically address bsAbs or fusion proteins for all validation experiments, the FDA guidance for validation of immunogenicity assays does specifically address the need for domain-specific characterization and states: "Examination of immune responses to therapeutic protein products with multiple functional domains such as bispecific antibodies may require development of multiple assays to measure immune responses to different domains of the molecules" (FDA 2019). As there may be multiple targets for the drug, the antibodies may have different specificities based on which portion of the drug structure they are generated in response to. In order to appropriately characterize the ADA response and based on appropriate risk assessment of the molecule, the immunogenicity assay for bsAbs may have additional tiers of confirmation for each functional domain of the protein. While screening and confirmatory of ADAs can occur using the entire bispecific or fusion protein as the critical reagent, additional characterization tiers will need custom reagents that represent each of the domains in order to determine how the ADA response is represented proportionally by each of the domains.

As validation aspects are considered, there are additional tweaks to experiments that may be required beyond the approach used for comparable mAb assay validation. As mentioned above, domain-specific cut points may need to be established for each functional domain of the protein. These domain-specific tiers may also require a positive control that is representative of the functional domain and not the whole molecule. If the positive control used for the whole drug screening and confirmatory assay is a pAb that represents the whole molecule, then a different domain-specific PC may be required for each of the domain-specific tiers. New PCs require that sensitivity be established in a similar fashion as the whole molecule PC, utilizing multiple dilution curves and analysts for each PC. The requirements for the domain-specific tiers enter into an area that is not clearly defined in most published guidance, but a general recommendation is that if a domain-specific cut point is established and domain-specific PCs are utilized, then control levels must also be adjusted for the PC. This includes evaluation of the domain-specific PC for appropriate LPC levels targeting the 1% fail rate. Additional validation testing may also be opted to be performed, such as selectivity and interference experiments with the domain-specific PCs. For target interference, if there are multiple targets of the molecule, both or all must be evaluated for potential interference in the assay.

Additional validation considerations can be summarized in the table below (Table 4.2).

Beyond the standard three tiers of analysis (screening/confirming/titer analysis) of the presence of ADAs toward bsAbs and fusion proteins, additional immunogenicity testing is required to determine if the detected antibody response contains neutralizing activity toward the drug. Neutralizing antibody (NAb) testing is

Table 4.2 Additional validation parameters for bispecific and fusion proteins

Validation parameter	Bispecific and fusion protein unique needs
Cut point determination	Additional domain-specific cut points may need to be established in confirmatory assay for characterization of ADA response
Sensitivity	If additional domain-specific cut points are determined with domain-specific PCs, then sensitivity for the domain-specific PC in screening, confirmatory, and domain-specific confirmatory may need to be reported
LPC/HPC determination	Appropriate control levels may need to be determined if a domain-specific PC is used in the domain-specific confirmatory tiers
Precision and acceptance criteria	Precision for domain-specific PC, including acceptance ranges and criteria, may be established
Selectivity and matrix interference	Selectivity and matrix interference testing may be evaluated with the domain-specific PCs following the same design as whole-molecule PC
Target interference	All targets of the whole drug molecule should be evaluated for potential interference in the assay

typically performed on samples that have screened and confirmed positive in previous tiers. While this generally means there are fewer samples to analyze, the impact of these results can be some of the most important data generated. NAb analysis may not be necessary for interpretation of toxicity data in preclinical studies, and therefore may not need to be implemented until clinical testing is required.

The nAb analysis may be cell-based or plate-based and often depends on multiple factors that can impact the decision on which path to pursue. If the therapeutic molecule is considered high risk for immunogenicity, as many bsAb and fusion proteins are, the preference is to use a cell-based assay that is based on the mechanism of action of the therapeutic. These assays can more accurately represent and measure the impact of nAbs in the therapeutic-specific pathway, but they are notoriously more complicated to develop, validate, and robustly implement. If the efficacy of the therapeutic is dependent upon binding of both targets simultaneously, then one cell-based nAb assay based on the mechanism of action of binding to either target may be acceptable. This functional assay would show the impact of nAbs on one target binding, and results could be interpreted to determine impact on PK/PD from this assay alone. If the bsAb or fusion protein does not rely on the synergistic effect of binding both targets, thus, both targets can bind independently and have independent efficacy for each binding action, and then two different NAb assays may be required for both functional domains (Shi et al. 2021).

Development of cell-based nAb assays uses the mechanism of action or pathway impacted by the drug to measure disruption in the presence of neutralizing antibodies, but this can often be difficult due to increased interference observed in assay development for bsAb and fusion proteins. In one example, a B-cell maturation antigen (BCMA)-CD3 bsAb neutralizing cell-based assay was developed based on the mechanism of the drug. This therapeutic binds to both BCMA on the surface of multiple myeloma cells and the CD3 on the surface of T-cells, utilizing a bridging and recruiting effort of immune cell responses. The cell-based assay measured T-cell activation via luciferase activity, which should be increased if the drug is able to work without nAb interference. Interference by soluble BCMA was observed during development and leads to increased false positives (reduction of luciferase not due to presence of nAb). Additional sample pretreatment steps using a bead-based approach were required to remove excess soluble BCMA from the sample matrix prior to analysis in the assay (Yang et al. 2019). This is just one example of the increased difficulty one may face in developing appropriate nAb assessments.

The overall bioanalytical strategy can be summarized as follows (Fig. 4.3):

4.2.2.2 Critical Reagent Generation

Critical reagent generation to support immunogenicity analysis of bsAbs and fusion proteins is often a complex process, as the reagents not only must be able to support the standard screening and confirmatory assays but also will need unique reagents for domain detection and antibody characterization assays. Additionally, the reagents, such as the positive controls, may need to be characterized to determine

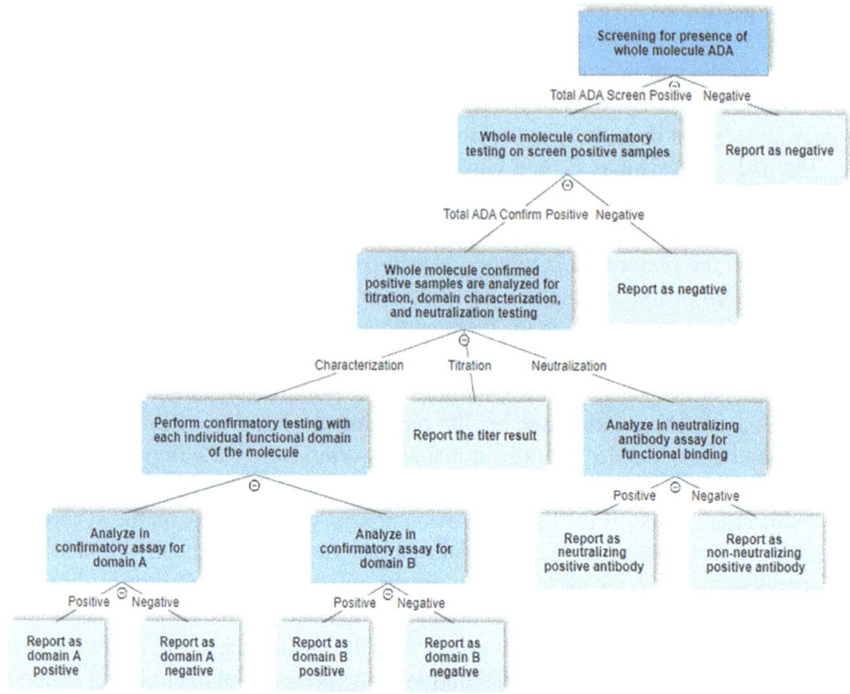

Fig. 4.3 Bispecific and fusion protein immunogenicity analysis flow chart. Note: This flow chart only includes neutralization with a functional binding assay. In the event a cell-based functional assay cannot be implemented, or if neutralization antibody information for each domain is desired, then two different domain-specific neutralization assays may be implemented similar to the domain characterization step

how much of the different domains of the molecule they represent and if the controls need to be supplemented or altered if they do not adequately represent the potential antibody response in patient samples.

Plate-based and cell-based immunogenicity assays may utilize common critical reagents, such as the positive controls, drug material, and target. These reagents may be labeled with additional tags to enable capturing and detection of the anti-drug antibodies in the assays, but labeling procedures for bsAbs and fusion proteins must take into account the variability in size of these different types of molecules and how the labeling needs to be incorporated for use in the assay. Arguably, the most important of all of these reagents is the positive control, as this represents the ADA responses expected to be measured in samples and needs to accurately reflect the crucial characteristics of the assay, such as sensitivity and drug tolerance. Positive controls, however, are considered reagents and not reference materials as true antibodies produced in vivo in response to therapeutic exposure cannot be used and instead surrogate PCs from animals are used to model the potential ADA response.

The 2019 FDA guidance on immunogenicity testing gives recommendations for positive control development. They recommend positive control antibodies should be generated by immunizing animals using the therapeutic protein product. After immunization, the future PC should be affinity purified. This approach should create a polyclonal antibody that is sufficient for representing ADA responses and allows for a better interpretation of sensitivity assessment results. The guidance also acknowledges that if a pAb cannot be created, individual mAbs or mAb combinations may be used, but it is recommended to engage with the agency for feedback if differing from standard approaches. As one can deduce, this approach will need to be altered and adapted for bispecifics and fusion proteins, as these molecules have structures that will directly impact how antibody controls can be generated due to multiple binding regions and targets. The positive control must represent the ADAs that may form in response to these treatments, and with multiple therapeutic targets being engaged by bsAbs and fusions, this may be hard to comprehensively generate a pAb in animal models.

Ideally, the PC used to represent a bispecific protein would have antibody binding characteristics for domains of the drug that are expected to generate immunogenic responses with high specificity. In order to bind different epitopes of an antigen, a polyclonal antibody should be used, as recommended in the guidance. To generate the pAb specific for ADA representation, the animal should be exposed to the whole drug molecule (all antigens) in order to ideally create a pAb control that can bind to multiple epitopes representing the different targets of the drug. While the representation may not be equal for all domains of the molecule, the PC should have some representation of all potential ADA responses. In the event a pAb cannot be generated to sufficiently represent the full ADA response covering all components of a bsAb or fusion protein, a panel of mAbs, or a "cocktail" approach of multiple mAbs, may be utilized as suggested in the guidance. In this approach, mAbs toward each domain of the bsAb or fusion molecule may be generated and used as a panel or combined together to represent all possible ADA responses. Whether a pAb or mAb panel is utilized may direct how certain validation experiments must be designed, as characteristics for each individual mAb component may need to be reported, such as sensitivity or drug tolerance. It should be noted that generation of these PC types is often time-consuming and must be performed in laboratories that specialize in pAb and mAb generation, which can impact timelines for assay development.

For NAb assays for bsAbs and fusion proteins, the positive control must exhibit neutralizing activity toward the drug binding or mechanism of action (depending on if using an LBA or cell-based assay). Other critical reagents may be shared with the ADA screening and confirmatory assays, but labeling ratios or concentrations in their respective assays will likely still need to be optimized.

4.2.2.3 Platforms

Appropriate platforms for bispecific and fusion analysis do not differ significantly from those discussed in previous chapters, but it is worth noting that immunogenicity rates can vary significantly based on platform and sensitivity achieved. For LBA, microplate readers such as SpectraMax and MSD (Meso Scale Discovery) are typical selections for plate-based assays as the instruments are commonly found across immunochemistry laboratories and are renowned for their robust results. More recent options for ADA analysis include Gyrolab technology, which historically had been utilized for predominantly PK analysis. The Gyrolab platform has the potential to offer increased sensitivity while utilizing a smaller sample aliquot and may eliminate target or matrix effects due to the microfluidic structure, both of which are problems that may be encountered in the common plate-based options. The platform may largely be determined by the quality and type of critical reagents that can be generated, and multiple options may need to be investigated to determine which produces the best sensitivity, drug tolerance, target tolerance, and avoidance of interfering factors. Transitioning from a standard bridging format for analysis to one of the more complicated methods discussed (bead based, solid-phase extractions, purifications, etc.) may also influence which instrument is most appropriate for readout.

For nAb analysis, the platform will be determined based on the decision to utilize a cell-based assay versus a plate-based assay. The options for these platforms do not differ significantly from what has been discussed in previous chapters, but the strategy for selection may vary based on the structure and need for additional immunogenicity information.

While LBA have been the standard approach for measuring immunogenicity, there has been an increasing trend in utilizing LCMS and LBA-LCMS hybrid assays for semi-quantitative analysis of ADA and immunogenicity responses. LCMS may be a superior choice when considering samples that may arrive in unusual biological matrices that do not perform well on plate-based methods, or if drug product is shown to be denatured during standard assay treatment during LBA analysis (such as denaturation in acid treatment). The LBA-LCMS hybrid approach is gaining in usefulness and popularity across the industry and is being employed for therapeutics that require isotyping results or do not fit with the traditional platforms for analysis. Additional discussion with regulatory agencies may be required to determine what approach will be most appropriate for the molecule and expected immunogenicity responses (Neubert et al. 2018).

4.2.2.4 Clinical Vs. Nonclinical Bioanalytical Strategies

Nonclinical bioanalysis does not necessarily follow the same regulatory analysis that is applied to clinical, human sample analysis. While ADA responses can be expected when dosing animals in nonclinical studies, their response rate is not often a good predictor of human immunogenicity responses. Based on this, the nonclinical immunogenicity approach for therapeutics may be minimal in comparison to the clinical approach and may be guided by a purpose-driven need instead of for safety or efficacy explanation purposes. The nonclinical approach for bsAbs and fusions proteins, however, often requires more in-depth scopes of their studies, and the need for bioanalytical testing is often higher.

Toxicity studies are necessary to determine safe first-in-human doses in clinical analysis, and it is recommended to use animal models that express the target in similar tissue distribution with representative similar activity of the drug, when possible. When considering there are multiple targets for the bsAbs and fusion proteins, this may complicate finding appropriate species to use for the toxicity studies. If an appropriate species cannot be identified, the toxicity programs may need to utilize surrogate cross-reactive molecules or transgenic animal models which express the human targets (Husain and Ellerman 2018).

Beyond toxicity studies, nonclinical studies may also be helpful or even necessary counterparts to interpreting and explaining nonclinical PK data. International Council for Harmonisation of Technical Requirements for Pharmaceuticals for Huamn Use (ICH) guideline S6 (2011) outlined the requirements for preclinical safety evaluation, which applies to bsAbs and fusion proteins. While this guidance offers some wider interpretation to the bioanalytical requirements than the comparable clinical guidance and allows for flexibility in whether or when preclinical testing should be implemented, it does explicitly state the nonclinical ADA testing should be implemented when there is: "1) evidence of altered PD activity; 2) unexpected changes in exposure in the absence of a PD marker; or 3) evidence of immune-mediated reactions (immune complex disease, vasculitis, anaphylaxis, etc." As noted in the introduction to bsAb structures, some bsAbs specifically utilize immune-mediated reactions (such as T-cell recruitment), which falls specifically into the category of explicitly required. It must be acknowledged, though, that preclinical testing is still limited due to the use of animal models and may still be poor indicators of adverse reactions or high immunogenicity levels, especially as these advanced bsAb and fusion molecules target human-specific pathways.

In addition to the ICH guideline S6, the FDA published a guidance specific to bispecific proteins in 2019 which included direct consideration of nonclinical analysis. While it did not clearly define nonclinical validation parameters for bioanalysis like sensitivity or cut point determination, it did outline the scope and guidelines for when to implement nonclinical testing. This guidance is further discussed in the "Regulatory Guidelines" section.

As for bioanalytical assay differences between nonclinical and clinical analysis, this will often be dependent upon the scope of the nonclinical study and what the data will be intended to support. In some cases, only screening data may be necessary in nonclinical studies, and so confirmatory cut points and domain characterization will not be performed.

4.2.2.5 Challenges in ADA Bioanalysis

The complexity of the molecules leads to many challenges in appropriately setting up the correct bioanalytical strategy. The strategy for one bsAb or fusion protein may not be the same for another, and many different factors must be considered. Structure, stability, reagent procurement, and assay sensitivity and specificity are just a few of the components which must be considered when designing the immunogenicity assays.

One unique challenge for bsAb and fusion proteins is the risk assessment for immunogenicity that does not apply to other modalities. Risk assessment for immunogenicity is considered before the assays have even been designed and influences the design, timeline, and implementation. Risk assessment starts with sequence analysis to determine if the sequence varies greatly from endogenous sequences and may be more likely to elicit an immunogenetic response than a comparable mAb therapeutic. After the sequence-based risk assessment, there are multiple other factors that must be considered for increased risk, such as product quality, and the product quality must also be evaluated for increased risk. Known product characteristics specific to bsAbs and fusion proteins that can impact immunogenicity include high-molecular-weight species and molecules that are likely to undergo biotransformation.

An article published in April 2020 by Broders et al. highlighted additional challenges in bioanalysis due to determining appropriate methods for domain characterization of bispecifics and fusions. As discussed in the bioanalytical strategy section, ADA characterization must occur after antibodies have been positively identified and confirmed. This characterization may occur through domain competition or domain detection assays. In these characterization tests, the biotherapeutic must be altered, usually through enzymatic cleavage, in order to allow for measurement of each domain or fragment of the molecule. The domains or fragments may be used as capture materials in the assay, targeting only antibodies that recognize only that specific fragment or domain. Alternatively, the fragments or domains can be used in a competition-type setting in addition to the labeled whole-drug material. The design of the domain characterization and detection assay may need to be tested in multiple formats in order to ensure the most sensitive and specific bioanalytical approach has been selected. These approaches are only appropriate for the larger bsAb structures (such as IgG-like bsAb) and fusion proteins, and smaller bsAbs with fragment structures likely cannot utilize these approaches as they cannot easily be broken down into stable subdomains (Broders et al. 2020).

The standard challenges in bioanalysis also apply to bsAbs and therapeutic proteins too. Considerations must be made for the disease states that will be evaluated during clinical studies, and the potential for matrix impact should be carefully evaluated before implementing the assay for reporting clinical data. Matrix effect can greatly sway cut point determinations, which are the crux for making an ADA assay result relevant. During assay development, the disease state populations should be evaluated to determine if they are comparable, or if individual disease states may be necessary. Alternatively, evaluating individuals and different matrix types may also reveal the assay format is inappropriate or subject to interference from matrix components, and other formats should be tested.

In general, the conventional straightforward bridging assay may not be sufficient for measuring Abs to bsAb and fusion proteins at an acceptable sensitivity level with minimal interference. One example of this is seen in bsAbs that contain two functional binding sites on the complementarity determining region (CDR) of each arm. These molecules are unable to utilize the bridge format for ADA detection as the arms of the ADAs will bind to only one bsAb molecule, instead of two independent molecules needed to form the characteristic "bridge" in the format (Shi et al. 2021).

High background and signal variation are often observed in IgG-based bsAbs due to target interference, and while evaluating different buffers, optimizing critical reagent concentrations, and increasing the MRD may help with reducing and controlling signal, these changes significantly impact the assay sensitivity and drug tolerance levels. Different formats and strategies may need to be tested, such as adding acid dissociations or utilizing other assay approaches, such as SPEAD (solid-phase extraction with acid dissociation), ACE (affinity capture elution assay), or PandA (precipitation and acid dissociation). If any of these approaches are evaluated, they will need to be specifically designed to work for the bsAb or fusion protein format and will likely require additional development activities to ensure the format has successfully reduced interference. The resounding message to bioanalytical scientists is that knowing and understanding the structure of the therapeutic can help guide the immunogenicity approach and potentially eliminate time spent on testing platforms that will not work for the structure.

4.3 Regulatory Guidelines

Many of the previously discussed regulatory guidance apply to fusion and bispecific protein bioanalysis. The approach for assay validation should follow the most recent industry guidances, such as the FDA's bioanalytical publications for PK and ADA assays, as well as the ICH M10 guidance, for example. While these documents may not specifically refer to the challenges presented by fusion and bispecific proteins, they should be utilized to guide development and validation activities as a foundation, while recognizing additional testing or validation activities may still be required.

In April 2019, the FDA issued an independent draft guidance pertaining to bispecific antibody development titled, "Bispecific Antibody Development Programs," in response to the increasing number of bsAbs in nonclinical and clinical trials. The guidance acknowledged that the current pathway for approval and evaluation of mAbs may not be sufficient for bsAb programs, and additional testing related to safety and efficacy may be warranted based on the dual-targeting nature and increased risk of immunogenicity. The guidance also states the need for scientific justification for use of a bsAb treatment through submission of data supporting decreased dosage with the bsAb, increased safety or efficacy, and unique mechanisms of action all compared to existing mAbs and combination of therapeutics currently available. This may include a request for a direct comparison of the bsAb to an approved monospecific product to the same target(s). Nonclinical and clinical data may be requested to determine the efficacy and safety of the bsAb compared to approved monotherapies, especially if there is noted concern for immunosuppression from treatment or if there is concern that only one of the bsAb's targets is the main result of efficacy.

The FDA guidance classifies bsAbs into two broad categories: bsAbs that bridge two target cells and bsAbs that do not bridge two target cells. The guidance includes further characterization guidelines such as determining whether both targets of the bsAb must be engaged at the same time for efficacy, determining affinity and binding kinetics for each target, and determining synergy of how the two bound targets interact. While most of the characterization and manufacturing processes may not differ from mAbs, there are additional attributes that will need to be characterized in ways that are specific to bsAbs, such as antigen specificity, affinity and on- and off-rates, avidity, potency, process-related impurities, formation of fragments or homodimers, stability, and half-life.

4.3.1 Nonclinical and Clinical Considerations

Both nonclinical and clinical studies should be tailored to the bsAb structure. Nonclinical studies are necessary for understanding the pharmacology and toxicology prior to initiating human dosing. While the scope of the nonclinical studies may be similar to the approach used for mAbs in regard to species selection, general toxicology, and reproductive toxicology, additional characterization and testing is required for each of the targets of the bsAbs. Each target should be evaluated for expression profiles and specificity to complete the understanding of the pharmacological and toxicological impact of the individual targeted domains in additional to the whole molecule. If the bsAb has an agonistic property, additional considerations may be needed for the dosage determination, and it is recommended to select doses using MABEL—minimally anticipated biological effect level. In general, the guidance highlights that the common approaches in nonclinical development for mAbs will be mostly applicable to bsAbs, with a few additional tests for characterization.

Similarly, for the clinical aspect of bsAbs, many of the expectations for mAbs will also apply. One area expected to be impacted is the pharmacodynamic (PD) assessment, as it must account for the binding of each target and the interaction or impact of one target binding on the other. The guidance recognizes that bsAbs may have biologically active and inactive forms, which may increase the need for multiple bioanalytical assay types to measure each form and distinguish between the different form types accordingly. Total antibody, bound, and unbound PK assays may be requested to characterize the drug form being quantitated in a sample. For ADA analysis, each domain must be considered, and knowledge of how immune response to one domain may impact the other domain is necessary. Multiple ADA assays, or multiple tiers within the same assay, may be necessary to understand the immune response and determine if there is a higher immune response to one specific domain versus the other.

4.4 Emerging and Novel Trends

With over 100 bsAb and fusion proteins in the pipelines across the industry, the methods for bioanalysis will likely continue to evolve as these therapeutics progress through preclinical and clinical trials. On the forefront of analysis is the pressing need to continue to develop sensitive and robust assays, which has seen to be difficult for bsAb and fusion antibodies based on their complex structures. New platforms should be investigated for both PK and ADA analysis, as they may allow for increased sensitivity without interference from soluble targets or free drug in sample matrices. Additionally, more consideration should be made for the usefulness of hybrid assays, such as LBA-LCMS hybrid assays for measurement of ADA responses to these complex molecules. LCMS for ADA analysis may help alleviate some of the challenges faced by these modalities, but there is still a need for regulatory and industry feedback on the best practices for validating acceptable assays for immunogenicity purposes.

As fusion proteins and bispecifics evolve and their structures become more complex through engineering, bioanalysis techniques will also need to improve. Trispecific proteins and new fusion structures are on the horizon, which will likely continue to face the same and new challenges presented to the existing molecules. Additionally, there is further investigation into combination therapies, such as combining CAR-T treatment with bsAb, which may require additional changes to the bioanalytical strategy. Some bsAb treatments have shown improved efficacy over the standard CAR-T treatments and, with the "off the shelf" availability, can be a promising path forward (Subklewe 2021). These updated treatment approaches may result in additional assays in the future, interference testing for combination therapies, and changes in requirement for sensitivity or tolerance.

Disclaimer Any opinions or forward-looking statements expressed are those of the authors and may not reflect views held by their employers (Janssen Biotherapeutics for Sanjeev Bhardwaj and EMD Serono Research & Development Institute, Inc./Merck KGaA, Darmstadt, Germany, for Kelly Covert and Hongmei Niu).

References

Broders O, Wessels U, Zadak M, Beckmann R, Stubenrach K. Novel bioanalytical method for the characterization of the immune response directed against a bispecific F(ab) fragment. Bioanalysis. 2020;12(8):509–17. https://doi.org/10.4155/bio-2020-0064.

Caliceti P. Pharmacokinetic and biodistribution properties of poly(ethylene glycol)–protein conjugates. Adv Drug Deliv Rev. 2003;55(10):1261–1277. https://doi.org/10.1016/S0169-409X(03)00108-X.

Cao M, Wang C, Chung WK, Motabar D, Wang J, Christian E, Lin S, Hunter A, Wang X, Liu D. Characterization and analysis of scFv-IgG bispecific antibody size variants. MAbs. 2018;10(8):1236–47. https://doi.org/10.1080/19420862.2018.1505398.

Chen X, Lee HF, Zaro JL, Shen WC. Effects of receptor binding on plasma half-life of bifunctional transferrin fusion proteins. Molecular Pharmaceutics 2011;8(2):457–465. https://doi.org/10.1021/mp1003064.

Conrad U, Plagmann I, Malchow S, Sack M, Floss DM, Kruglov AA, Nedospasov SA, Rose-John S, Jürgen, J. ELPylated anti-human TNF therapeutic single-domain antibodies for prevention of lethal septic shock. Plant Biotechnol J. 2011;9(1):22–31. https://doi.org/10.1111/j.1467-7652.2010.00523.x.

Datta-Mannan A, Croy JE, Schirtzinger L, Torgerson S, Breyer M, Wroblewski VJ. Aberrant bispecific antibody pharmacokinetics linked to liver sinusoidal endothelium clearance mechanism in cynomolgus monkeys. MAbs. 2016;8(5):969–82. https://doi.org/10.1080/19420862.2016.1178435.

Fares FA, Suganuma N, Nishimori K, LaPolt PS, Hsueh AJ, Boime I. Design of a long-acting follitropin agonist by fusing the C-terminal sequence of the chorionic gonadotropin beta subunit to the follitropin beta subunit. Proc Natl Acad Sci. 1992;89(10):4304–4308. https://doi.org/10.1073/pnas.89.10.4304.

FDA Guidance for industry: Bispecific Antibody Development Programs, DHHS FDA Docket No. FDA-2019-D-0621; 2019.

Fischer SK, Yang J, Anand B, Cowan K, Hendricks R, Li J, Nakamura G, Song A. (2014) The assay design used for measurement of therapeutic antibody concentrations can affect pharmacokinetic parameters. mAbs. 2012;4(5):623–631. https://doi.org/10.4161/mabs.20814.

Giorgetti J, Beck A, Leize-Wagner E, François YN. Combination of intact, middle-up and bottom-up levels to characterize 7 therapeutic monoclonal antibodies by capillary electrophoresis—mass spectrometry. J Pharm Biomed Anal. 2020;182. https://doi.org/10.1016/j.jpba.2020.113107.

Haraya K, Tachibana T, Igawa T. Improvement of pharmacokinetic properties of therapeutic antibodies by antibody engineering. Drug Metab Pharmacokinet. 2019;34(1):25–41. https://doi.org/10.1016/j.dmpk.2018.10.003.

Husain B, Ellerman D. Expanding the boundaries of biotherapeutics with bispecific antibodies. BioDrugs. 2018;32(5):441–64. https://doi.org/10.1007/s40259-018-0299-9.

International Council for Harmonisation of Technical Requirements for Pharmaceuticals for Human Use. ICH harmonized tripartite guideline on preclinical safety evaluation of biotechnology-derived pharmaceuticals S6(R1), Step 4 Version; 2011.

Iwamoto N, Shimada T. Regulated LC-MS/MS bioanalysis technology for therapeutic antibodies and Fc-fusion proteins using structure-indicated approach. Drug Metab Pharmacokinet. 2019;34(1):19–24. https://doi.org/10.1016/j.dmpk.2018.10.002.

Iwamoto N, Yokoyama K, Takanashi M, Yonezawa A, Matsubara K, Shimada T. Application of nSMOL coupled with LC-MS bioanalysis for monitoring the fc-fusion biopharmaceuticals etanercept and abatacept in human serum. Pharmacol Res Perspect. 2018;6(4):1–10. https://doi.org/10.1002/prp2.422.

Jeremy Woods R, Xie MH, Kreudenstein TSV, Ng GY, Dixit SB. LC-MS characterization and purity assessment of a prototype bispecific antibody. MAbs. 2013;5(5):711–22. https://doi.org/10.4161/mabs.25488.

Kaur S, Liu L, Cortes DF, Shao J, Jenkins R, Mylott WR, Xu K. Validation of a biotherapeutic immunoaffinity-LC-MS/MS assay in monkey serum: "plug-and-play" across seven molecules. Bioanalysis. 2016;8(15):1565–77. https://doi.org/10.4155/bio-2016-0117.

Kaur S, Xu K, Saad OM, Dere RC, Carrasco-Triguero M. Bioanalytical assay strategies for the development of antibody–drug conjugate biotherapeutics. Bioanalysis. 2013;5(2):201–226. https://doi.org/10.4155/bio.12.299.

Kellie JF, Pannullo KE, Li Y, Fraley K, Mayer A, Sychterz CJ, Szapacs ME, Karlinsey MZ. Antibody subunit LC-MS analysis for pharmacokinetic and biotransformation determination from in-life studies for complex biotherapeutics. Anal Chem. 2020;92(12):8268–77. https://doi.org/10.1021/acs.analchem.0c00520.

Kernstock R, Sperinde G, Finco D, Davis R, Montgomery D. Clinical immunogenicity risk assessment strategy for a low risk monoclonal antibody. AAPS J. 2020;22(3):60. Published 2020 Mar 17. https://doi.org/10.1208/s12248-020-00440-5.

Knight T, Callaghan MU. The role of emicizumab, a bispecific factor IXa- and factor X-directed antibody, for the prevention of bleeding episodes in patients with hemophilia A. Ther Adv Hematol. 2018;9(10):319–34. https://doi.org/10.1177/2040620718799997. Published 2018 Oct 10

Kontermann R, Brinkmann U. Bispecific antibodies. Drug Discov Today. 2015;20(7):838–47. https://doi.org/10.1016/j.drudis.2015.02.008.

Labrijn AF, Janmaat ML, Reichert JM, Parren PWHI. Bispecific antibodies: a mechanistic review of the pipeline. Nat Rev Drug Discov. 2019;18(8):585–608. https://doi.org/10.1038/s41573-019-0028-1.

Li H, Ortiz R, Tran L, Hall M, Spahr C, Walker K, Laudemann J, Miller S, Salimi-Moosavi H, Lee JW. General LC-MS/MS method approach to quantify therapeutic monoclonal antibodies using a common whole antibody internal standard with application to preclinical studies. Anal Chem. 2012;84(3):1267–73. https://doi.org/10.1021/ac202792n.

Ma M, Colletti K, Yang TY, Leung S, Pederson S, Hottenstein CS, Xu X, Warrino D, Wakshull E. Bioanalytical challenges and unique considerations to support pharmacokinetic characterization of bispecific biotherapeutics. Bioanalysis. 2019;11(5):427–35. https://doi.org/10.4155/bio-2018-0146.

Manikwar P, Mulagapati SHR, Kasturirangan S, Moez K, Rainey GJ, Lobo B. Characterization of a novel bispecific antibody with improved conformational and chemical stability. J Pharm Sci. 2020;109(1):220–32. https://doi.org/10.1016/j.xphs.2019.06.025.

Mullard A. FDA approves first bispecific. Nat Rev Drug Discov. 2015;14:7. https://doi.org/10.1038/nrd4531.

Neubert H, Olah T, Lee A, Fraser S, Dodge R, Laterza O, Szapacs M, Alley SC, Saad OM, Amur S, Chen L, Cherry E, Cho SJ, Cludts I, Donato LD, Edmison A, Ferrari L, Garofolo F, Haidar S, Hopper S, Hottenstein S, Ishii-Watabe A, Kassim S, Kurki P, Lima Santos GM, Miscoria G, Palandra J, Pedras-Vasconcelos J, Piccoli S, Rogstad S, Saito Y, Savoie N, Sikorski T, Spitz S, Staelens L, Verthelyi D, Vinter S, Wadhwa M, Wang YM, Welink J, Weng N, Whale E, Woolf E, Wu J, Yan H, Yu H, Zhou S. 2018 White Paper on Recent Issues in Bioanalysis: focus on immunogenicity assays by hybrid LBA/LCMS and regulatory feedback (part 2—PK, PD & ADA assays by hybrid LBA/LCMS & regulatory agencies' inputs on bioanalysis, biomarkers and immunogenicity). Bioanalysis. 2018;10(23):1897–917. https://doi.org/10.4155/bio-2018-0285. Epub 2018 Nov 29

Roopenian DC, Shreeram, S. FcRn: the neonatal Fc receptor comes of age. Nat Rev Immunol. 2007;7(9):715–725. https://doi.org/10.1038/nri2155.

Runcie K, Budman DR, John V, et al. Bi-specific and tri-specific antibodies-the next big thing in solid tumor therapeutics. Mol Med. 2018;24:50. https://doi.org/10.1186/s10020-018-0051-4.

Sawyer WS, Srikumar N, Carver J, Chu PY, Shen A, Xu A, Williams AJ, Spiess C, Wu C, Liu Y, Tran JC. High-throughput antibody screening from complex matrices using intact protein electrospray mass spectrometry. Proc Natl Acad Sci U S A. 2020;117(18):9851–6. https://doi.org/10.1073/pnas.1917383117.

Schellenberger V, Wang C, Geething NC, Spink BJ, Campbell A, To W, Scholle MD, Yin Y, Yao Y, Bogin O, Cleland JL, Silverman J, Stemmer WPC. A recombinant polypeptide extends the in vivo half-life of peptides and proteins in a tunable manner. Nature Biotechnology 2009;27(12):1186–1190. https://doi.org/10.1038/nbt.1588.

Stevenson L, Kelley M, Gorovits B, Kingsley C, Myler H, Österlund K, Muruganandam A, Minamide Y, Dominguez M. Large molecule specific assay operation: recommendation for best practices and harmonization from the global bioanalysis consortium harmonization team. AAPS J. 2014;16(1):83–88. https://doi.org/10.1208/s12248-013-9542-y.

Shi J, Chen X, Diao J, et al. Bioanalysis in the age of new drug modalities. AAPS J. 2021;23:64. https://doi.org/10.1208/s12248-021-00594-w.

Subklewe M. BiTEs better than CAR T cells. Blood Adv. 2021;5(2):607–12. https://doi.org/10.1182/bloodadvances.2020001792.

Swann PG, Shapiro MA (2011) Regulatory considerations for development of bioanalytical assays for biotechnology products. Bioanalysis. 2011;3(6):597–603. https://doi.org/10.4155/bio.11.27.

Tibbitts J, Canter D, Graff R, Smith A, Khawli LA. Key factors influencing ADME properties of therapeutic proteins: a need for ADME characterization in drug discovery and development. MAbs. 2016;8(2):229–45. https://doi.org/10.1080/19420862.2015.1115937.

Trivedi A, Stienen S, Zhu M, Li H, Yuraszeck T, Gibbs J, Heath T, Loberg R, Kasichayanula S. Clinical pharmacology and translational aspects of bispecific antibodies. Clin Transl Sci. 2017;10:147–62. https://doi.org/10.1111/cts.12459.

Voronin K, Allentoff AJ, Bonacorsi SJ, Mapelli C, Gong SX, Lee V, Riexinger D, Sanghvi N, Jiang H, Zeng J. Synthesis of a stable isotopically labeled universal surrogate peptide for use as an internal standard in LC-MS/MS bioanalysis of human IgG and Fc-fusion protein drug candidates. J Label Compd Radiopharm. 2014;57(9):579–83. https://doi.org/10.1002/jlcr.3218.

Wang C, Vemulapalli B, Cao M, Gadre D, Wang J, Hunter A, Wang X, Liu D. A systematic approach for analysis and characterization of mispairing in bispecific antibodies with asymmetric architecture. MAbs. 2018;10(8):1226–35. https://doi.org/10.1080/19420862.2018.1511198.

Wang, YMC, Jawa V, Ma M. Immunogenicity and PK/PD evaluation in biotherapeutic drug development: scientific considerations for bioanalytical methods and data analysis. Bioanalysis 2014;6(1):79–87.

Wang Y, Li X, Liu YH, Richardson D, Li H, Shameem M, Yang X. Simultaneous monitoring of oxidation, deamidation, isomerization, and glycosylation of monoclonal antibodies by liquid chromatography-mass spectrometry method with ultrafast tryptic digestion. MAbs. 2016;8(8):1477–86. https://doi.org/10.1080/19420862.2016.1226715.

Yang W, Luong M, Guadiz C, Zhang M, Gorovits B. Addressing soluble target interference in the development of a functional assay for the detection of neutralizing antibodies against a BCMA-CD3 bispecific antibody. J Immunol Methods. 2019;474 https://doi.org/10.1016/j.jim.2019.112642.

Chapter 5
Gene Therapy and Cell Therapy: Bioanalytical Challenges and Practical Solutions

Darshana Jani, Ramakrishna Boyanapalli, and Liching Cao

Abstract Clinical development of gene therapy (GTx) and cell therapy (CTx) modalities involve extensive and challenging immunogenicity and pharmacokinetic assessments. The challenges associated with immunogenicity assessment are attributed to the unique delivery systems and complex nature of the immune responses against these multicomponent drug modalities. Many challenges for the immunogenicity assays are largely rooted in the critical reagents and analytical technology availability. Since viral vectors are commonly used for gene therapy, bioanalytical method development focusing on the assessment of pre-existing and drug-induced humoral anti-drug antibody (ADA) and cellular responses against adeno-associated virus (AAV) capsid and transgene expressed proteins require analytical technologies like the enzyme-linked immunospot (ELISpot) and flow cytometry beyond the conventional ligand binding assays (LBA) and liquid chromatography coupled mass spectrometry (LC-MS) platforms. Likewise, the bioanalytical strategy for chimeric antigen receptor (CAR) protein on CAR-T-cells also demands unique analytical platforms such as polymerase chain reaction (PCR). The regulatory expectations on bioanalytical technologies used for gene and cell therapy modalities continue to evolve. An industry consensus and regulatory guidance will help navigate through bioanalytical data and associated pharmacokinetics and immunogenicity data interpretation for these novel modalities.

Keywords Gene therapy · Cell therapy · Pharmacokinetics · Immunogenicity · ELISpot · PCR · Flow · Oligonucleotides · LC-MS · Ligand binding assays

D. Jani (✉)
Moderna Inc, Cambridge, MA, USA
e-mail: Darshana.jani@modernatx.com

R. Boyanapalli
Wave Life Sciences, Cambridge, MA, USA

L. Cao
Sangamo Therapeutics, Brisbane, CA, USA

S. Kumar (ed.), *An Introduction to Bioanalysis of Biopharmaceuticals*, AAPS Advances in the Pharmaceutical Sciences Series 57,
https://doi.org/10.1007/978-3-030-97193-9_5

5.1 Introduction

Gene therapy (GTx) and cell therapy (CTx) modalities require unique bioanalytical (BA) considerations for immunogenicity assessments against multiple components of the biotherapeutic agent (gene or engineered cells) and determination of multiple non-conventional analytes for pharmacokinetics (PK) and/or pharmacodynamics (PD) endpoints. The comprehensive immunogenicity and PK/PD assessment when combined with critical reagent generation leads to significant planning of time and resources. In addition, the design and execution of BA assays and reporting of the resulting data have yet to be standardized unlike conventional small-molecule and large-molecule biotherapeutics where regulatory expectations are clearly defined for their PK and immunogenicity assessment using traditional bioanalytical platforms (liquid chromatography coupled mass spectrometry [LC-MS], ligand binding assay [LBA]). The regulatory guidance and/or industry consensus on bioanalytical technologies such as flow cytometry, enzyme-linked immunospot (ELISPot), and polymerase chain reaction (PCR), typically used for GTx and CTx modalities, continues to evolve. A combination of traditional standard BA platforms and new creative approaches are thus needed to generate confidence in BA assay performance and data for these emerging modalities. This chapter discusses the challenges of assessing BA risks and developing appropriate testing strategies for immunogenicity and PK analysis for this promising class of modalities. Biomarker assays are outside the scope of this chapter. The goal of this chapter is to highlight the BA strategies and their successful application for clinical studies.

5.2 What Is Gene and Cell Therapy?

According to the National Library of Medicine, GTx is described as "Gene therapy is an experimental technique that uses genes to treat or prevent disease. In the future, this technique may allow doctors to treat a disorder by inserting a gene into a patient's cells instead of using drugs or surgery."

As per European Medical Agency (EMA) guidelines published in 2015, gene therapy consists of a delivery system that contains a "genetic construct engineered to express a specific therapeutic sequence or protein responsible for the regulation, repair, addition, or deletion of a genetic sequence" (Mingozzi and High 2013). In common terms, gene therapy is a technique that delivers genetic materials, such as therapeutic nucleic acid polymers, to a patient's cells with the goal of treating or curing a disease at the DNA or RNA level and targeting diseases that do not respond to traditional drug therapies or intent to provide a potential one-time treatment. The therapy can be used to reduce levels of a disease-causing protein, increase production of disease-fighting proteins, or to produce new or modified proteins. In contrast to gene therapy, cell therapy infuses/transplants *whole cells* into the patient. The cells may originate from the patient (autologous cells) or a donor (allogeneic cells).

The third type of cell therapy is mostly based on the cell type used for the therapeutic indications such as somatic cells, immortalized, etc., which are explained in detail in the PK assessment section of this chapter (Mount et al. 2015). In the case of chimeric antigen receptor (CAR) T-cell therapy, immune cells such as T-cells are modified to express a receptor on the surface that recognizes antigens on targeted cells to provide the treatment.

For successful delivery of GTx, two different gene delivery systems are typically used: viral- and non-viral-based vectors. In viral vector-based GTx, a healthy DNA sequence is inserted into the vector, called vector genome or payload (Kaiser 2020), which is then injected into the patient. Like a regular virus, the vector "infects" the patient's cells and thus deposits the healthy DNA sequence into the nucleus of the cell. The patient's cells begin to generate normal proteins. Rare diseases, those that are caused by a monogenic single mutation, are considered an ideal target for gene therapy. Most GTx use viral vector-based delivery systems (High and Roncarolo 2019), with the most common being adeno-associated virus (AAV), although lentivirus and retrovirus are also frequently used. Non-viral vector-based delivery systems for GTx include lipid nanoparticles (LNP) and microbes. Although useful, nonspecific cytotoxicity has been observed with cationic liposomes. Microbes used as vectors are mainly bacteria, and the process is known as bactofection (Palffy et al. 2006). After decades of research, LNP-based programs have shown clinical utility where a whole mRNA capable of expressing protein in vivo would be delivered in combination with ionized cationic lipids as LPS delivery mechanism in anticancer, antiviral, and certain systemic intracellular therapeutics (Kulkarni et al.

Table 5.1 List of FDA-approved GTx and CTx products

Drug	Therapy	Indication	Approval year
LUXTURNA (voretigene neparvovec-rzyl)	AAV9 gene therapy	Retinal dystrophy associated with bi-allelic RPE65 mutation	2017
ZOLGENSMA® (onasemnogene abeparvovec-xioi)	AAV9 gene therapy	Spinal muscular atrophy (SMA) for <2 years old associated with missing or nonworking survival motor neuron 1 (SMN1)	2019
Yescarta™ (axicabtagene ciloleucel):CD 19	CD 19 T-cell therapy	Immunotherapy medicine to treat large B-cell lymphoma in adults	2017
KYMRIAH® (tisagenlecleucel): CD19	CD19 T-cell therapy	Autologous T-cell immunotherapy indicated for young adults aged <25 years with B-cell precursor acute lymphoblastic leukemia	2018
TECARTUS™ (brexucabtagene autoleucel): CD19	CD19 T-cell therapy	Adult patients diagnosed with mantle cell lymphoma (MCL)	2020
Abecma (idecabtagene vicleucel):BCMA	BCMA-T-cell therapy	Adult patients diagnosed with multiple myeloma	2021

Gene and cell-based therapies approved by FDA in last 5 years

2018; Gomez-Aguado et al. 2020). One clinical trial initiated for bactofections in 2002 with attenuated *Salmonella typhimurium* in melanoma and renal carcinoma patients (Kramer et al. 2018). However, due to lack of efficacy, the trial was terminated after phase I. In 2016, FDA provided recommendations for microbes as GTx vectors (MGTVs) (FDA 2016).

Due to the prevalence of viral-based gene therapies, this chapter will mainly focus on bioanalysis of viral vector-based gene therapy drug development.

Currently, in the USA, there are two approved GTx and three approved CTx products as listed in Table 5.1.

5.3 Bioanalysis

Bioanalysis of GTx and CTx consists of immunogenicity, PK, and biomarker evaluations. Bioanalytical challenges for GTx can vary depending on the type of vector, genetic material (transgene) properties, the target organs, and the administration system. To develop an effective immunogenicity assessment strategy for GTx, both humoral and cellular immune responses to the transgene and vector are considered (Fig. 5.1). Like conventional multi-domain large-molecule biotherapeutics, humoral immune response for GTx is also assessed as anti-drug antibody (ADA) response mounted after exposure to different functional domains of biotherapeutic agent. Additionally, assessment of pre-existing ADA response to AAV vectors and to prior protein-based therapies is also essential.

Immune response evaluations are essential to correlate with safety and efficacy of the therapeutic. Measurements of the biotherapeutic agent (DNA or RNA) for GTx are performed using technologies like quantitative polymerase chain reaction

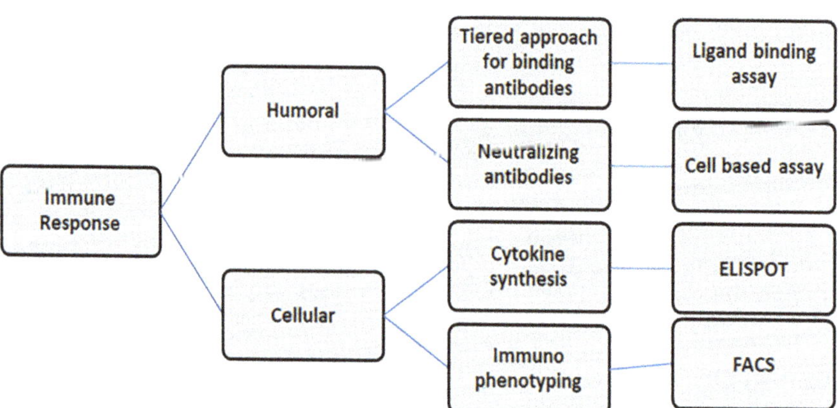

Fig. 5.1 Gene therapy: Example of immunogenicity assessment strategy for GTx: Evaluation includes both humoral and cellular immune responses to the transgene and vector

(qPCR), hybrid enzyme-linked immunosorbent assay (ELISA), positron emission tomography (PET) imaging, measurements of the viral or nanoparticle vector (e.g., qPCR, LC-MS), and production of a missing protein (using LBA). These measurements would not likely correspond to conventional PK endpoints on a discrete analyte but may involve concentration-time measurements of cells utilizing flow cytometry or molecular approaches or testing how the immune system is being augmented and its impact on the associated safety. The main objective of measuring the various analytes is to understand exposure vs efficacy correlation of gene product, biodistribution (Zolgensma product label: Biodistribution was evaluated in two patients who died after the infusion), and its shedding into various body matrices such as blood, saliva, and urine.

In case of cell therapies, CAR-Ts are "living" drugs and are designed to expand in vivo. Approaches to evaluate the humoral immunogenicity toward extracellular domain of receptor is similar to that of large molecules using bridging immunoassay following multi-tiered strategy. Importantly, unlike large molecules, assessment of cellular immunogenicity by assays such as ELISpot is needed. The assessment strategy for evaluating anti-CAR-T antibodies is similar to what is used for protein therapeutics. It is important to remember that immunogenicity sampling times is important since lymphodepletion will impact immunogenicity. Sampling times used in tisagenlecleucel provides good guidance for such evaluation (Mueller et al. 2018; Awasthi et al. 2020).

CAR-T safety and efficacy are related to CAR-T expansion within the body. The maximum CAR-T levels are generally achieved within ~2 weeks of infusion. The majority of exposure is observed within the first month following infusion; however, it is required that the percentage of CAR-Ts needs to be tracked over time since CAR-Ts may persist for years. Typically, cells are directly quantified using flow cytometry, or CAR-T transgene expression is measured using qPCR or digital droplet PCR (ddPCR) assay.

Overall, some of the desirable features of GTx and CTx BA assays include:

- Assay is robust and reproducible irrespective of whether the samples are fresh or frozen.
- Assay works with genetically diverse population.
- Assay uses the least amount of cells/clinical sample material, as possible.
- Assay can accommodate high-throughput testing (hundreds of samples in the later stages of a clinical trial).
- Assay can be standardized to allow inter-study comparisons for multi-center or international clinical trials.

In addition to various technology platforms, laboratories must have a predetermined protocol for how, when, and in what quantity biospecimens must be collected for the clinical study, as well as a plan to store, transport, and test them.

The following sections describe immunogenicity and PK evaluations used for GTx and CTx and may provide solutions to specific challenges faced by bioanalysts.

5.4 Immunogenicity Evaluation

GTx and CTx offer treatment for patients with inherited or acquired diseases. Clinical development of these novel biotherapeutic modalities involves extensive and challenging immunogenicity assessment. The immunogenicity assessment plan is designed to evaluate potential immune response to multiple components of the treatment, including response to delivery vehicles such as recombinant adeno-associated virus (rAAV) vectors, DNA or RNA encoded transgenes, and the transgene expressed protein for GTx. For CTx, immune response to the chimeric antigen receptor on CAR-T should be evaluated.

This immune response assessment not only requires evaluation of post-treatment ADA response but also potential pre-existing neutralizing antibodies (NAbs) to rAAV vectors to determine enrollment eligibility in clinical studies. Additionally, assessment of pre-existing ADA response to protein-based therapies (such as clotting factors) and enzyme replacement therapy (ERT) is also essential as biotherapeutic treatment may present heightened immunogenicity risk and reduce the efficacy. Unique challenges associated with ADA assays for gene therapies are discussed using rAAV GTx and CAR-T CTx as examples.

Serological studies show that individuals develop humoral immunity against AAV early in life due to exposure to the wild-type virus (Mingozzi and High 2013). The first gene therapy clinical study using rAAV2 vectors to deliver the factor-IX gene through the systemic administration showed that low pre-existing anti-AAV2 NAb titers completely neutralized doses of vector and negatively impacted clinical efficacy (Manno et al. 2006; Mingozzi and High 2013). Protein replacement or ERT has become the standard of care for diseases caused by a specific protein deficiency, and patients often develop high levels of persistent antibodies to such protein products based on their genetic background and other factors which may lead to loss of efficacy (Melton et al. 2017; Pan et al. 2017; Solomon and Muro 2017; Goh and Ng 2018; Mauhin et al. 2018). NAb status to AAV vectors and the presence of inhibitor to hemophilia factor treatment, therefore, have been used as inclusion/exclusion criteria for patient enrollment to GTx clinical studies to ensure potential treatment benefit. In addition to induced humoral responses, GTx treatments can also induce cellular responses and immune-mediated hepatotoxicity (Manno et al. 2006; Pien et al. 2009; Nathwani et al. 2011a, b; Masat et al. 2013; Colella et al. 2018; Gorovits and Koren 2019).

From a BA perspective, challenges of immunogenicity assessment specific to GTx are attributed to the complex nature of the immune responses against these multiple-component drugs (e.g., AAV vector, transgene) and the extent of assessment needed at different stages of clinical development. In contrast, large-molecule biotherapeutics only need assessment of post-treatment ADA response to the protein drug.

Similar to large-molecule biotherapeutics, many factors contribute to immunogenicity of a GTx product including the manufacturing process (impurity and aggregate), drug administration (route, dose, and frequency), patient characteristics (immune responsiveness and mutation status or defective gene), and inherent properties of the biotherapeutic molecule (protein sequence and mechanism of action). All these contribute to immunogenicity risk assessment factors that need to be considered when designing a bioanalytical strategy for immunogenicity evaluation. Due to the diverse scope of possible immune responses that can result from GTx and CTx treatments, this chapter focuses on assessment of pre-existing and drug-induced humoral ADA and cellular responses against AAV vector capsid and transgene expressed proteins or the CAR protein on CAR-T-cells. Considerations for assay development strategies, challenges, and solutions, as well as regulatory aspects of BA method development, are discussed and applicable to both GTx and CTx immunogenicity assessments.

5.4.1 Binding Antibody Assays

5.4.1.1 Strategy

Immunogenicity assessment is critical in GTx drug development because viral delivery vehicles such as rAAV vectors, gene editing tools such as the bacterial Cas9 enzyme, DNA or RNA encoded transgene, as well as transgene-expressed proteins may cause ADA responses which reduce clinical efficacy (Masat et al. 2013; Mingozzi and High 2013; Simhadri et al. 2018; Charlesworth et al. 2019; Gorovits and Koren 2019). Conventional ADA methodologies such as ELISA or Meso Scale Discovery (MSD) chemiluminescence platform used for biotherapeutic proteins are applicable to GTx (Jani et al. 2015; Partridge et al. 2016). The choice of assay format (bridging or direct binding) and analytical platform depends on individual preferences and methods best suited to address the question of immune responses elicited from the therapy. Strategies evaluating humoral immune responses induced by different drug components should be specifically tailored to the study drug using risk-based assessment (Shankar et al. 2008, Agency 2017, Ma et al. 2017, FDA 2019a, b). An example is mRNA therapies, where the induction of immunogenicity against mRNA is considered a low risk; therefore, routine monitoring may not be necessary. However, the formation of antibodies against the mRNA encoded protein following repeat dosing should be monitored routinely in clinical studies (Millipore n.d., Ma et al. 2017, Kowalski et al. 2019).

The multi-tiered immunogenicity testing strategy commonly used for large-molecule biotherapeutics employs a sequential approach including screening, confirmatory, titration, and neutralization assays, followed by further characterization of isotyping and epitope mapping, if needed (Shankar et al. 2008; Agency 2017; FDA 2019a, b). The first three tiers are based on binding assays and are followed by NAb assays which are typically functional assays. Though this multi-tiered approach provides a good framework for assessment, the sequence of the tiered approach may not be suitable for immunogenicity assessments of GTx products. For clinical study enrollment, pre-existing immunogenicity to rAAV capsid is commonly evaluated using a cell-based rAAV-fluorescent reporter transduction inhibition assay (Meliani et al. 2015) or total antibody assay. The NAb assay for screening may not have a confirmatory step for practical reasons. To assess pre- and post-GTx ADA for rare disease clinical studies, the small sample size may benefit from an operation and cost perspective by combining screening and titration together as top-tier analysis followed by a confirmatory assay.

5.4.1.2 Challenges and Solutions

Common issues BA scientists face during binding ADA assay development include generating multiple critical reagents, setting up assay ADA cut points in presence of pre-existing antibodies, increasing assay sensitivity, and drug tolerance. Availability of critical reagents such as transgene-expressed proteins and positive control (PC) antibodies is the main challenge for ADA assay development. For large-molecule biotherapeutics, reagents such as ERT products or antibodies to the ERT protein are readily available, well characterized, and in ample quantity for use as in assays or immunogens for PC antibody generation. GTx and CTx studies partially rely on commercially available recombinant proteins, such as recombinant enzymes as surrogates for transgene expressed proteins in vivo. In some instances, this lack of commercially available reagents requires custom production of single-chain variable fragment (scFv) domains of the chimeric antigen receptor for CAR-T-cells, genome editing components, or novel proteins specific to the therapy. Hence, the selection and characterization of a surrogate protein becomes critical as the recombinant protein may not fully represent the protein expressed through GTx treatment in vivo for detecting relevant drug-induced ADA response. Different expression systems for generating the recombinant protein such as mouse NSO myeloma cells versus Chinese hamster ovary (CHO) cells may impact post-translational modifications (Buttel et al. 2011; Goh and Ng 2018). This could impact assay specificity and lead to false-positive or false-negative responses, for example, nonspecific heterophilic interactions. In ADA assays, these reagents may require conjugation with tags (such as biotin) or detecting agents (such as ruthenium (II) tris bipyridine-(4-methylsulphone)). Post-labeling characterization is essential as high molar incorporation ratio of tag or detecting agent to protein may impact the protein structure and function, especially when the expressed transgene protein is an active enzyme. Enzyme activity tests should be implemented post conjugation to confirm that the conformation of the enzyme and function have not been altered. To ensure assay consistency and long-term use of the ADA method, storage buffer and conditions for these reagents should be carefully evaluated as well (Kubiak et al. 2016).

Positive control antibody generation, a general limitation to method development for ADA assay, presents another layer of complexity for GTx immunogenicity method development. Surrogate PC antibodies may be generated using the recombinant protein to immunize animals or by phage display methods. Like ADA assays for biotherapeutic proteins, purified monoclonal and polyclonal antibodies are acceptable for use as positive controls (Agency 2017; FDA 2019a, b). However, the validity of using these reagents needs to be characterized extensively in the assay development phase to ensure specificity as both the transgene expressed protein and PC antibody are surrogate reagents. Since the recombinant transgene protein needs to be produced before a PC antibody can be made, generation and characterization of both reagents may take 6 months or more. The lengthy timeline for reagent generation can be a challenge as GTx and CTx studies are often on a fast track for regulatory approval.

Drug tolerance is another major challenge in ADA assay development because the presence of high concentrations of expressed protein in serum may interfere with ADA binding to assay reagents which may result in false-negative results. For biotherapeutic proteins, it is recommended that blood samples be obtained after an appropriate washout period to minimize the effect of drug tolerance (FDA 2019a, b). Due to continuous and long-term expression of the protein from GTx treatment, transgene product-free samples cannot be obtained. Though a drug tolerant assay provides better sensitivity, the clinical relevance of positive results from a highly sensitive assay is initially unknown for GTx. To ensure achievement of the required sensitivity per regulatory guidance, different sample pre-treatment approaches (acid dissociation, heat pre-treatment, solid-phase extraction with acid dissociation (SPEAD), affinity capture elution (ACE), etc.) may be needed to detect ADA in the presence of the targeted drug level (Bourdage et al. 2007; Smith et al. 2007).

5.4.2 *Neutralizing Antibody (NAb) Assay*

5.4.2.1 Strategy

Assessing the immunogenicity risk of NAbs is critical during the clinical phase of drug development. The testing method selected to assess neutralizing potential for ADA should be based on the mechanism of action of the drug (Agency 2017, FDA 2019a, b). For GTx, examples of commonly used NAb assays are cell-based transduction inhibition, cell proliferation, protein uptake inhibition, and enzyme activity inhibition assays. A cell-based transduction inhibition assay is commonly used in clinical studies as the assay to evaluate pre-existing immunity to the viral capsid AAV for enrollment purposes (Fig. 5.2) (Falese et al. 2017). This remains the current assay of choice since the method can detect NAbs as well as other factors in blood that could potentially impact AAV transduction. However, total antibody (TAb) binding assays are emerging as an alternative approach (Fig. 5.3). Concordance between TAb and NAb assays was observed in the majority of tested disease and healthy donor samples with TAb assays appearing to correlate with transgene expression better than NAb assays in nonclinical studies (Falese et al. 2017; BioMarin 2019; Long et al. 2019). Another approach such as AAV-mediated liver transduction using a mouse neutralization activity model was used as a tool to understand the impact of AAV NAbs, but these assays can be cumbersome, time-consuming, not predictive, and difficult to implement as a screening test for clinical

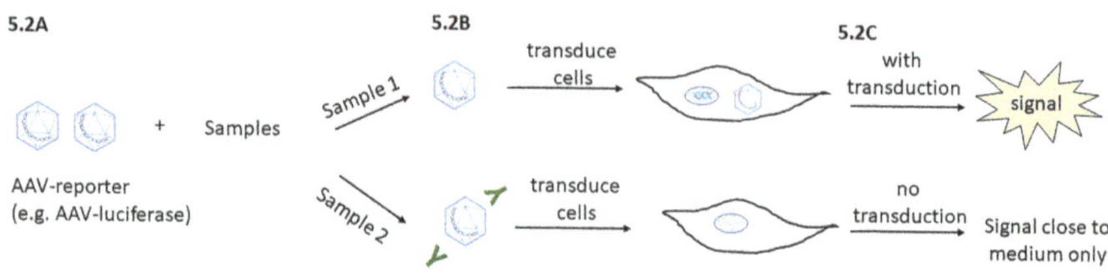

Fig. 5.2 Reporter transduction inhibition assay. (**a**) AAV-reporter (e.g., AAV-luciferase) and test samples (e.g., sample 1 and sample 2) are preincubated. (**b**) Cells are transduced overnight with AAV-reporter/sample mixture. (**c**) Evaluate luciferase expression the following day. Sample with neutralizing antibody or neutralizing factors to AAV will reduce transduction and hence results in lower luminescence. All samples are normalized to negative control to assess inhibition

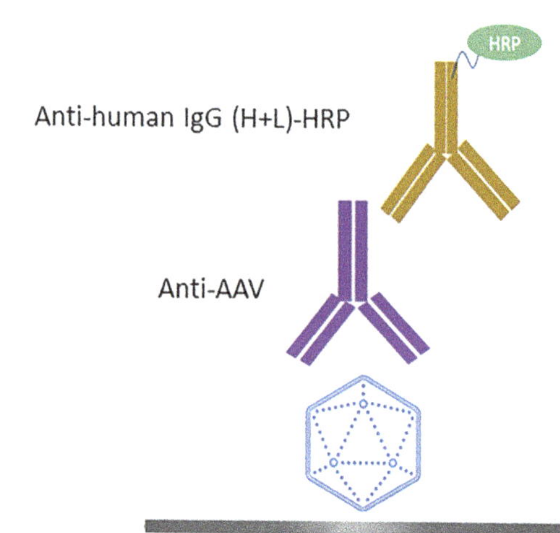

Fig. 5.3 Total antibody binding assay. An example of a total antibody binding assay for AAV by ELISA format. AAV capsid is coated on the bottom of microtiter plates. Bound anti-AAV antibodies are detected by secondary anti-human IgG antibody conjugated to HRP

studies (Scallan et al. 2006; Kruzik et al. 2019; Long et al. 2019). An in vitro NAb assay that quantitates surface-bound rAAV vector genome copy number in cells by RT-qPCR is being explored as an alternative method as well (Guo et al. 2019).

For GTx studies, where transgene expressed protein is an active enzyme, it is essential to understand the mechanism of action of the enzyme when designing the NAb assay strategies. More than one assay may be needed to cover different neutralization pathways and mechanisms including antibodies that inhibit enzyme activity and antibodies that prevent cell surface receptor binding and subsequent internalization of the enzyme. Both types of NAbs have been reported to reduce efficacy. For the latter type of NAbs, various approaches such as inhibition of cell-surface receptor binding and neutralization of cellular uptake have been used to evaluate the impact of ADA (Melton et al. 2017; Cheung et al. 2018). For CAR-T-cells, NAb activity can be assessed by measuring the potential reduction of CAR binding to its molecular target in a cellular environment or reduced cytotoxicity response to target tumor cell (Gorovits and Koren 2019).

5.4.2.2 Challenges and Solutions

NAb assays and TAb binding assays share challenges including critical reagents, pre-existing antibodies, assay sensitivity, and drug tolerance. For cell-based assays, critical reagents such as cell line selection, accuracy of AAV reporter concentration, and full vs empty capsid ratio determined by the manufacturer are important, especially for NAb assays to evaluate pre-existing immunity to rAAV. Several cell lines have been used in the literature to evaluate the impact of pre-existing immunity to AAV on transduction efficiency. The vector transduction efficiency is cell line dependent and needs to be evaluated as poor in vitro transduction efficiency requires a greater multiplicity of infection (MOI) which results in lower assay sensitivity and, therefore, underestimation of AAV Nabs (Meliani et al. 2015). AAV reporter concentration is typically measured by qPCR which has higher assay variability and can impact MOI of NAb assay. Alternative approach by ddPCR might be a better approach for AAV reporter concentration determination.

For GTx, where the expressed transgene is an enzyme, different assays and PCs may be needed to characterize the confirmed ADA, which includes assessment of ADA binding to proteins involved in cellular uptake or enzyme catalytic domains (Concolino et al. 2018). Commercial or custom recombinant proteins, approved protein-based treatments, or synthetic peptides located at the enzyme active site or antigen uptake site can be used as immunogens to produce surrogate PC antibodies. It is possible, however, that none of these approaches can generate neutralizing PC with enough sensitivity. When all these approaches are exhausted, donor sera with pre-existing NAbs or NAb positive sera to the expressed transgene from clinical trial subjects post treatment can be used as a positive control. In order to define assay sensitivity, NAbs need to be purified from sera using an affinity column. If the amount of sera is limited or affinity purification approach is not feasible, immunoglobulin can be extracted using protein A/G purification, and a NAb concentration can be assigned by calibration-free concentration analysis (CFCA) using Biacore (Pol et al. 2016). Several GTx studies are evaluating lysosomal storage enzymes that are active in acidic conditions. The impact of the assay conditions such as optimal pH for ADA assay to allow the detection of low-affinity antibodies should be evaluated. Similar to binding antibody assays, due to constant expression of the protein from GTx treatment, transgene product-free samples cannot be obtained during the treatment phase of the trial, and drug tolerance will need to be evaluated.

For pre-existing immunity to the AAV capsid, a wide range of seropositivity has been reported for different serotypes, but the biological and potential clinical relevance of these data remains elusive. This is largely due to different assay methodologies, minimum required dilution, assay reagents, and populations studied (BioMarin 2019; Kruzik et al. 2019; Majowicz et al. 2019). There are two common approaches for determining the cutoff threshold to evaluate seropositivity, including utilization of titer at half maximum inhibition (IC50) or cut point determination per immunogenicity guidance (Shankar et al. 2008; Meliani et al. 2015; Agency 2017; Falese et al. 2017; FDA 2019a, b; Long et al. 2019; Majowicz et al. 2019). Currently, there is no consensus on assay strategies and approaches to set cutoff threshold. As described by Kruzik et al. (2019), the lack of standardized assays and accurate description of NAb methods can result in controversial outcomes (Kruzik et al. 2019). This calls for a unified standard methodology for each serotype to allow comparison of work by different sponsors and simplification of the AAV screening process. Patients with low NAb titer may still have treatment benefit, especially in high AAV dose cohorts, but the impact of NAbs on efficacy needs to be determined clinically.

Pre-existing NAbs elicited from prior therapies such as protein replacement and ERT may negatively impact efficacy of GTx products and add complexity to assay development. Current hemophilia AAV GTx studies exclude subjects with inhibitor targeting the clotting factor using a well-established Bethesda test because the inhibitor is known to affect patient safety and treatment efficacy (Doshi and Arruda 2018). However, standardized commercial tests to evaluate NAbs elicited from various ERT treatments are not available, and a screening test may be needed for subject enrollment in GTx studies with pre-existing NAbs to ERT. Current immunogenicity guidelines use 5% false-positive rate for screening and 1% for confirmatory assays to determine antibody incidence post treatment, while gene therapy immunogenicity assays used for pre-screening may need different considerations. The current statistical approach used to define antibody incidence post treatment may not be suitable for enrollment criterion establishment.

Additionally, difficulties in obtaining ERT-treated patient samples to evaluate the impact of pre-existing NAb to GTx products and drug-naïve patient samples to establish disease-specific cut point for enrollment in rare disease clinical studies add another hurdle for assay development.

5.4.3 *Cellular Immunity*

5.4.3.1 Strategy

The importance of T-cell-mediated immunity to AAV capsid was initially evidenced in the first AAV GTx using an AAV2 vector to deliver the factor 9 gene. Two subjects had transient elevation of liver enzymes and loss of FIX transgene expression due to capsid-specific T-cell response immune targeting transduced hepatocytes (Manno et al. 2006). The ELISPot platform is most widely used for evaluating T-cell responses to drug delivery vectors such as AAV capsid and transgene proteins (Fig. 5.4). This functional T-cell assay utilizes freshly isolated or thawed cryopreserved peripheral blood mononuclear cells (PBMC) from subjects. PBMCs are incubated with antigens (peptide pools) to allow for antigen-specific T-cell activation. The induction of cytokines such as IFN-γ secreted from activated T-cells is the common readout and involves counting spots of stained IFN-γ. Each individual spot represents an individual IFN-γ-secreting cell. This assay is sensitive and allows for the detection of antigen-specific immune cells in low frequencies. Strategies for method development, validation, and supporting multi-center clinical trials and assay harmonization are reviewed by Tary-Lehmann et al. (2008), and Britten et al.

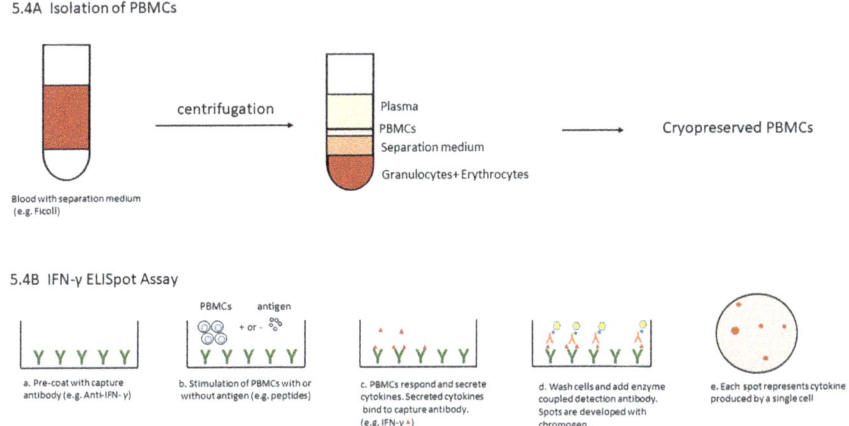

Fig. 5.4 Schematic diagram of PBMC isolation (**a**) and IFN-γ ELISpot assay (**b**). (**a**) PBMCs are isolated from whole blood and cryopreserved. (**b**) (a) Plate is precoated with capture antibody (e.g., anti-IFN-γ); (b) PBMCs are stimulated with and without antigens (e.g., peptides); (c) PBMCs respond and secrete cytokines. Secreted cytokines bind to capture antibody (e.g., IFN-γ); (d) wash cells and add enzyme coupled detection antibody; spots are developed with chromogen; (e) each spot represents cytokine produced by a single cell

(2013). Strategies to evaluate cellular immune response to cell therapies such as CAR-T-cells for oncology indications are well described by Gorovits and Koren (2019).

5.4.3.2 Challenges and Solutions

The ELISpot assay is complex, and it is challenging to compare data across different laboratories due to dissimilarities of protocols, reagents, instruments, spot counting and reporting methods, individual preferences, and personnel training. The assays are time sensitive, and PBMCs need to be isolated within 24 hours and cryopreserved for future testing. Critical factors such as blood collection tube and anticoagulant, possible stabilizer used, and the pre-isolation blood storage temperature and time should be carefully evaluated. Quality of the cryopreserved PBMCs and purity of synthetic peptides (AAV or transgene) are two critical components for ELISpot assay development. Cell numbers and peptide pool concentrations need to be optimized to maximize the signal-to-noise ratio to avoid erroneous or ambiguous results (Millipore). Moreover, ELISpot assay tends to have higher assay variability due to complexity of the assay. Intra- and inter-assay precision evaluation needs to be evaluated using positive-control donor samples to better understand assay variation.

Though ELISpot assay can be standardized, challenges still remain. ELISpot assays cannot distinguish between $CD4^+$ and $CD8^+$ responses without depleting or enriching one of these subsets. Because high numbers of cells are required, this is not practical to obtain a sufficient volume of blood from small infants. T-cell responses may be underestimated as only IFN-γ is measured and this does not provide information on T-cell ability to proliferate or their activation/exhaustion status (Ertl and High 2017). Multicolor flow cytometry which allows the analysis of up to 16 parameters on any given cell (Mittag and Tarnok 2011) and mass cytometry which can analyze greater than 40 different parameters on a single cell (Leipold et al. 2015) can also be used to track T-cell responses in rAAV gene transfer recipients as alternative approaches (Ertl and High 2017).

Developing an ELISpot assay for GTx leads to a product-specific assay for cellular monitoring due to specific vectors and biotherapeutic transgenes. This poses a challenge for the GTx community as different approaches may be used across a variety of clinical products and studies and the results may not be comparable across institutions (Britten et al. 2013). Recent harmonization of the ELISpot protocol (Britten et al. 2013) has helped narrow the gap and allows for more precise measurement. Adaptation of ELISpot to clinical use will require a rigorous quality program to track assay performance.

5.4.4 Regulatory Perspective on Immunogenicity Assessment for GTx and CTx

GTx and CTx offer much promise for treatment of numerous inherited or acquired diseases. This significant growth of GTx and CDx clinical trials prompted a number of recent FDA guidelines such as the guidance on Human Gene Therapy for Rare Diseases and disease-specific guidance for hemophilia, retinal disorders, and Fabry disease (FDA 2019a, b, FDA 2020a, b, c). These FDA regulatory documents have touched upon the key parameters to assess safety and clinical efficacy but lack details and directions on how to approach assay development and validation. Currently, many BA laboratories follow the general method validation guidance for immunogenicity assays for biotherapeutic proteins (Agency 2017; FDA 2019a, b). The guidance does provide a good framework but is not applicable for all types of immunogenicity assays for GTx and CTx studies due to the complex nature of immune response to the complex drug. Additionally, a consensus on assay strategies and approaches for setting enrollment criterion is needed when an ADA assay is used for screening subjects with pre-existing antibodies. Current immunogenicity guidelines recommend using a 5% false error rate for screening and 1% for confirmatory assays when accessing ADA development post-treatment, while GTx immunogenicity assays may be needed to establish enrollment criterion. In most cases, NAb or TAb assays are being used for enrollment, and 1% or 0.1% false error rates may be more suitable for this application in order to reduce the chance of excluding false-positive patients.

Different approaches and methodologies may create substantial divergence between BA data and data interpretation (Ma et al. 2017). An alignment in best practices is needed for GTx and CTx immunogenicity assessment. Having a well-established set of industry practices and regulatory guidance in the future will help align the BA approaches for evaluating immune responses elicited by GTx and CTx products.

5.5 Pharmacokinetic Evaluations

PK parameters for small-molecule or large-molecule drugs include drug exposure, clearance, and half-life in nonclinical and clinical studies. Concurrently wherever feasible, certain PD endpoints are also considered for PK/PD evaluations at nonclinical and clinical stages of a program. However, these traditional PK study designs and evaluation are generally not feasible for GTx and CTx products. For GTx and CTx, this translates into understanding the initial dose, clearance, biodistribution, and duration of cell bioavailability. Similar to small-molecule and biologics therapies, in GTx and CTx modalities, PK endpoints predominantly involve the effect and body response to the biotherapies. In addition, PD evaluations take precedence at various stages of a GTx or CTx program. The PK/PD evaluations in all the modalities can be similar endpoints (Ma et al. 2017; Ma et al. 2021). Traditional BA platforms such as LC-MS and ligand binding assays may not provide desired specificity and sensitivity in pharmacokinetic (PK) measurements. Other platforms such as molecular methods and flow cytometry should be explored to quantify the administered drug using PK assays in nonclinical or clinical studies. In this chapter, these non-traditional methodologies required for PK measurements in GTx and CTx (specifically chimeric antigen receptor T CTx [CAR-T]) are explained. In addition, alternate approaches to assess biodistribution and exposure of the biotherapeutic agents are explained in this chapter.

5.5.1 PK Assessment Strategies for GTx

In GTx modality, desired recombinant genetic material or transgene is delivered into the host using various vectors such as viruses and microorganisms. The vectors that are predominantly used are viruses. Adeno-associated viruses (AAV), with linear single-strand recombinant DNA, are widely used viruses for packaging the transgene since complete genomic sequence is available for various AAV vectors and a well understood host-virus interaction, thorough understanding of virus transduction and replication, and relatively mild innate and adaptive immune responses are shown when injected into the host (Samulski and Muzyczka 2014; Kuranda et al. 2018; Shirley et al. 2020). Other vectors such as LNPs and bactofections were previously discussed. However, due to limited use of LNPs and MGTVs and similarities with the parameters used in overall PK assessments, the strategies discussed in this chapter will be using viral vectors.

During nonclinical stages of drug development, known amounts of virus will be dosed in animal models or toxicology species to understand PK and PD parameters, and these studies facilitate identification of therapeutic window and therapeutic index of a drug (Savva 2019). The amount of virus would be denoted in viral genomes per volume and quantified using analytical methods involving polymerase chain reaction (PCR) (Weltner et al. 2012). A drug molecule will be an intact virus that is engineered to deliver a given payload. For PK assessment, viral payload should be measured mostly in blood with time after dosing and reported as viral genomes per volume. The bioanalysis in nonclinical studies the payload will be measured in systemic biofluids or biofluids from disease regions, where possible, for PK assessment. In addition, the drug is also measured in target tissues for drug exposure in target engagement or safety studies. Based on the therapeutic area, tissue matrix might vary. These bioanalyses assess biodistribution and safety of a drug. For ADME studies, samples can be biofluids such as plasma, urine, cerebrospinal fluid, and target tissue or peripheral tissue. As a bioanalyst, PK assays should be developed in each matrix and qualified or validated based on study compliance requirements (Steinmetz and Spack 2009).

In clinical studies for viral vector-based gene therapy programs, PK and biodistribution-based exposure assessment in clinical studies is a challenging task due to limitations of obtaining relevant samples, especially tissues for conducting these assessments. PK measurements of transgene or vector itself, performed in blood samples using various BA tools, may not provide complete details. So, in these cases, measuring the transgene product, pharmacodynamics (PD) will be used as a main assessment to study GTx drug concentration. An advantage to obtain exposure and distribution using PD assessment would be that the same datapoint can also be used as a biomarker. However, additional biomarker assessments and any respective assays, which are specific to target engagement or proof of principle, will not be discussed in this chapter. For PD assessment, various methodologies are available, and a relevant method should be chosen to develop as a quantitative assay (Stevenson et al. 2018). In addition, the data obtained during nonclinical development in animals may provide useful data for initial calculations. However, nonclinical to clinical translation of PK and exposure data may not be accurate.

5.5.2 Strategic Considerations for CTx

Regulatory guidance for nonclinical studies of cell therapies was initially established in 1998 (FDA 1998) and updated in 2013. Majority of the cell therapies can be grouped into three major groups (Mount et al. 2015).

5.5.2.1 Somatic Cell Therapies

Cells used in this treatment will be isolated from a donor and purified before administering to a patient. These cells will not undergo any molecular change before administering (Demchuk et al. 2016). Examples for these cells are stem cells from various origins from a donor such as hematopoietic, skin, mesenchymal, umbilical cord, etc. Certain T-cell therapies might also be part of somatic cells if the cells did not require any technology change. However, these therapies are not very common compared to the cells with any form of technology change. PK assessment of these cell therapies involves whole-cell-based assays that are specific to the cells used for CTx (Kakkanaiah et al. 2018). Typically, flow cytometry is used for whole-cell PK assessment. Interference from patient cells might lead to challenges to specifically identify the therapeutic cell over patient cells. In case identifiable surface markers specific to the therapeutic cells, which are not seen in patient blood, are expressed by the CTx, these markers will be used in PK assessments by using fluorescent tagged antibodies specific to the surface markers. When incubated with the fluorescent antibodies, the cells with the surface markers will be fluorescent and enumerated using flow cytometry. In case the surface markers are not expressed in the CTx developed, alternate methods are used such as qPCR to absolutely quantitate the cells with the transgene (Yang and Doddareddy 2020; Ma et al. 2021).

5.5.2.2 Cell Therapies with Ex Vivo Modification Using Viral Technologies

Cells used in these treatments are predominantly various types of T-cells, such as CAR-T, natural T killer cell therapies, etc. T-cells from a patient will be isolated from their whole blood and engineered ex vivo with recombinant receptors using viruses such as lentivirus (Chen et al. 2019). These engineered CAR-T-cells should be measured with time after dosing the drug as a PK marker. Various molecular technologies which can quantitatively measure the transgene will be used. If the transgene results in a cell surface protein, which is specific to the therapeutic cell, the engineered surface protein can also be used as a marker for quantification (Milone and Bhoj 2018). Similar to somatic cell therapies, based on cell engineering, either flow cytometry or qPCR methods would be used to quantitate the cells.

5.5.2.3 Immortalized CTx

The cells used in this therapy are mostly stem cell types of somatic origin. These cells will be immortalized by incorporating a transgene using a viral transduction. The immortalized cells can help treat patients with prolonged injury of an organ such as the spinal cord (Santiago-Toledo et al. 2019). Since the transgene is not naturally seen, it will be quantitated for PK assessment purposes using qPCR methods to quantitate the transgene.

During clinical studies, challenges similar to GTx are expected in all types of CTx. To summarize, the main differences in PK assays between the therapies are the type of tools used to engineer a gene or cell. In GTx the drug to be quantified in a bio-sample will be the vector used, which may not have been exposed to the study subject before the study. However, in cell therapies, the cells used as therapies might have originated from the patient or from a donor. Same cell type as the therapy cell can be seen in the subject resulting in specificity issues while developing a PK assay (Milone and Bhoj 2018; Chen et al. 2019). To overcome these challenges, additional tools will be needed. Below section will have details on choices of methodologies, which can be used in GTx and CTx therapies.

5.5.3 *Bioanalytical Methods for PK Evaluations of GTx and CTx Therapies*

Bioanalysis of these novel modalities such as GTx and CTx can be achieved using long-standing technologies such as quantitative (qPCR) or quantitative real-time polymerase chain reaction (qRT-PCR) and flow cytometry and next-generation technologies such as digital PCR (dPCR) and branched DNA (bDNA) technologies (FDA 2016; Gupta et al. 2017; Kakkanaiah et al. 2018; Welink et al. 2018; Piccoli et al. 2019). These age-old technologies are widely implemented in PK assessments in GTx and cell therapies. However, next-generation technologies (Hu and Huang 2020) and ultrasensitive assays such as targeted sequencing (Breton et al. 2020), single-cell sequencing (Santeramo et al. 2020), and immuno-PCR (Dovgan et al. 2019) have also shown tremendous promise in quantitating drug molecules. Additional new technologies such as Single Molecule Array (SIMOA®) (Li et al. 2008), Single Molecule Counting (Torchinsky and Ebenstein 2016), and CRISPR/Cas9 (Roidos et al. 2020)-based novel approaches are pushed as potential technologies for quantitating nucleic acids. However, their usage is not thoroughly studied. So, these technologies will not be discussed in this chapter. The technologies, which are compatible for bioanalysis of drug molecules in GTx and CTx modalities, especially CAR-T, are itemized in Table 5.2.

Table 5.2 Biotherapeutic modalities and their respective platforms to develop PK assays (FDA 2016, Gupta et al. 2017, Kakkanaiah et al. 2018, Welink et al. 2018, Piccoli et al. 2019)

Biotherapeutic modality	PK/PD assay platform	Target measured
GTx	qPCR dPCR bDNA ELISA	Transgene Transgene Transgene Gene product (PD)
CAR-T	qPCR dPCR bDNA Flow cytometry ELISA	Transgene Transgene Transgene Whole cell (drug) Gene product (PD)

In GTx modalities, the target engaging component is always nucleic acids packaged as payload in virus. The form of nucleic acid used will be depending on the vector used to deliver in these therapies. To measure dosed GTx or CTx therapeutics, molecular techniques should be a preferred approach over traditional bioanalysis. Every virus molecule has a copy of the nucleic acid molecule intended to be delivered. Quantification of vectors can be more reliable in molecular techniques widely used to quantify nucleic acids, while in CAR-T therapy, the transgene used can be quantified using similar technologies as GTx. PK assays for almost all types of cell therapeutics can be performed on the molecular and cell-based assays listed in GTx section except somatic cell therapies with no technology alternation with a transgene.

These molecular techniques listed in Table 5.2 have varying sensitivities and require different instruments and reagents. An assay development scientist can choose a platform based on the available instruments and resources. Overall, all the molecular techniques listed have better sensitivities than immunoassays. However, these methods require expensive equipment and reagents. Various platforms used for PK assessment in GTx are listed in Fig. 5.5.

5.5.4 PK Assay Development Strategies

At present, there are no specific guidance available on PK assay validations using PCR-based methods. There are continuous discussions with industries involved in GTx and CTx modalities and regulatory agencies, and very soon some type of guidance may be released from EMA or FDA (Yang and Doddareddy 2020; Ma et al.

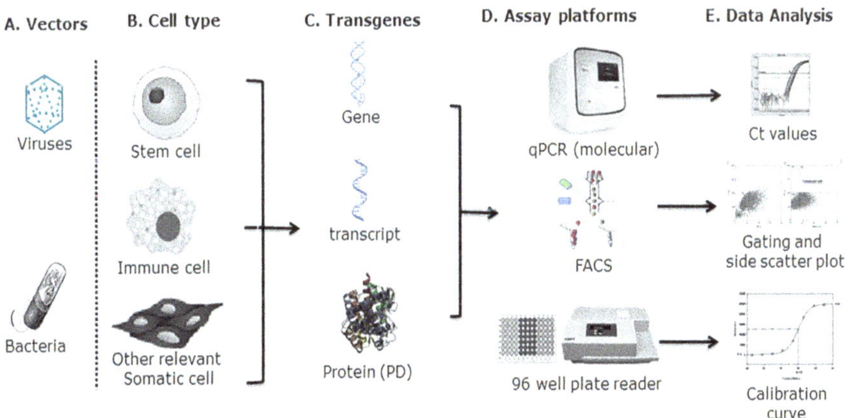

Fig. 5.5 Platforms used for PK assessment in GTx. Schematic diagram of a PK assays from drug component of a GTx and CTx delivery vehicle to the options of instruments available to choose for quantification of the drug

2021). Usually, PK assay development should be carried out based on the recommendations provided by regulatory guidance such as the FDA document finalized in 2018 (FDA 2018). However, the assays described in the guidance is specific to LC-MS and ligand binding assays. Currently, there are white papers available, which should be thoroughly read and understood before assay development is started (Gupta et al. 2017; Kakkanaiah et al. 2018; Stevenson et al. 2018; Welink et al. 2018; Piccoli et al. 2019; Ma et al. 2021). Assays should be prepared in the same matrix as the sample types expected for PK measurements. All the assays used for crucial nonclinical studies and clinical samples should be validated using good laboratory practices (GLP). The parameters listed in the white papers cited in Table 5.3 should also be considered while validating the assays.

While qPCR and qRT-PCR assays are developed for PK assessments, it is important to use absolute quantification methods. Standards required for absolute quantification should be prepared by using transgene with known copy numbers of the genome. Purified transgene from vectors or cells should be used to prepare standards and quality controls (QCs) for the assay. If viral transgene is not available, synthetic transgene reagents should be used to prepare standards and QCs (Table 5.3).

When flow cytometry-based PK assays are developed for cell-based therapies, fluorescent markers or probes specific to the engineered cells should be used. The standards and QCs should be prepared with the cells with known cell counts from analytical-grade assays and in the same sample matrix. These engineered cells without fluorescent probes should be used for gating. Same gating should be used for all standards and QCs (Table 5.3). All parameters listed in the white papers (Gupta et al. 2017; Kakkanaiah et al. 2018; Stevenson et al. 2018; Welink et al. 2018; Piccoli et al. 2019) and PK guidance (FDA 2018) should be validated before proceeding with sample testing. In nonclinical studies, similar methods can be used to make assays in specific sample matrices to run tissue exposure and biodistribution, viral shedding for GTx, and urine PK sample testing.

In viral vector-based gene therapy programs, PK-based exposure assessment in clinical studies is a challenging task due to limitations of obtaining relevant samples, especially tissues to do these measurements. Another approach to obtain exposure and distribution in these situations would be to measure the transgene product (pharmacodynamics, PD) using assays such as RT-PCR or immunoassays developed specifically for the PD (Piccoli et al. 2019). If the PD assays used for PK assessment are also used to test mode of action and proof of principle studies, the same PD assay can also be used as a biomarker assay in relevant sample types. Overall, novel gene modulating therapeutics such as GTx and CTx provide opportunities to expand the horizon of applying diverse scientific methodologies in the field of bioanalytics. These individual modalities have different approaches to deliver recombinant genetic material or a payload.

Table 5.3 Key considerations for selecting PCR vs FLOW technology for Pk assessment for GTx and CTx

Attribute	qPCR/qRT-PCR technology (Yang and Doddareddy 2020; Ma et al. 2021)	FLOW Technology (Yamamoto et al. 2020; Mfarrej et al. 2021)
Analyte	• Transgene DNA (e.g., AAV vectors-based gene therapy) • cDNA of transgene (e.g., lentivirus vector-based CTx) • Product of transgene (mRNA) – PD readout • For standards synthetic transgenes can also be used	• Target T-cells in CTx (autologous or allogenic cells) • In GTx, transgene is expressed in immune cells • Gating for the target cells and same gating should be used across the assay validation
Matrix tested	• Biofluids (plasma, serum, CSF, feces, and urine) • Target tissue (liver, kidney, spleen, etc.)	• Predominantly immune cells from preclinical and clinical studies
Sample preparation	• Direct phenol – chloroform or bead extraction • Nucleic acids extracting columns are also available • Tissues need additional homogenization step	• Hemolyzing the sample to remove RBC and followed by washing to remove cell debris • Washing the cells in the dispersing medium and incubating with surface or intracellular target binding antibodies with fluorescent tags
Reagents	Lysis buffer for tissues, primers, probes, Taq polymerase	Antibodies, assay-specific buffers, fluorescence conjugated (antibodies)
Assay parameters		
Sensitivity	LLOQ for transgene is 20 copies in 400 ng gDNA	LLOQ was 10 cells/100 μL
Assay variability	Accuracy and precision by two separate analysts on multiple days	
Assay linearity	Minimum transgene needed for PCR amplification, R^2 value, and slope needed to be tested	R^2 value and slope needed to be tested
Selectivity	>10 individual subjects' whole blood nucleic acids should be tested	Not usually tested since cells are isolated from biomatrix and washed before incubated with the surface specific antibody reagents. So, patient matrix interference is minimal due to the processing steps
Specificity	Transgene amplified product size and sequencing	Antibody specificity to the cell type or expressed transgene should be tested
Stability	Long term, short term, and freeze thaw stability of transgene in the sample	

5.5.5 *Discussion*

Novel gene modulating biotherapeutics such as GTx and CTx provide opportunities to expand the horizon of applying diverse scientific methodologies in the field of bioanalysis. These individual modalities have different approaches to deliver recombinant genetic material. Though individual bioanalytical assays for GTx and CTx therapy are not more complex than traditional BA assays, there are modality-specific factors, e.g., study design, patient and site enrollment, duration of testing, multiple components of the treatment, etc., that need to be considered. All these factors collectively make overall GTx and CTx projects more complex and challenging. Unclear regulatory guidance for gene biotherapeutic products could also hinder the BA method development/validation if various issues, questions, and options arise. Further complications could arise when bioanalysis in multiple matrices within tight timelines is needed.

Immunogenicity is an essential part of safety assessments in clinical development of gene therapies. Both humoral (binding/neutralizing antibodies) and cellular immune responses are tested for gene therapy programs. Traditional risk assessment can be carefully applied based on criteria such as patient population, route of administration, and vector utilized in the study; additional criteria such as manufacturing process (impurity and aggregate) and inherent properties of the biotherapeutic molecule (protein sequence and mechanism of action) should also be considered. This assessment requires evaluation of not only post-treatment ADA response but also potential pre-existing NAbs to the delivery system (e.g., AAV) to determine enrollment eligibility in clinical studies as well as pre-existing ADA response to prior protein-based treatment. Assessment of cellular immunity using ELISpot type of unique assay formats is also an important part of strategy providing understanding of efficacy and toxicities. Appropriate reagents such as procuring positive control for such assays is time-consuming and can be expensive. Many publications are available on how to standardize ELISpot methods; however its validation is complex and requires regulatory attention. In late 2000, there was a harmonization effort to standardize the technique which provides useful tips on method parameters to consider (Janetzki et al. 2008) during assay validation and life cycle management. Sample collection is demanding since samples can be collected at multiple time points; however, relevance of sample assessment at multiple time points, biological value of the resulting data, and clinical correlation with multiple endpoints should be carefully considered. Prior consultation with health authorities may provide further insight into what is needed to address safety and overall tolerability. The samples can be stored frozen if risk assessment is ongoing, and long-term follow up is needed

For drug-disposition studies, GTx and CTx studies do not follow traditional PK endpoints for dose response and dose selection. Instead, drug effect in the pharmacological and toxicological species is assessed as PD effect and body response to the biotherapeutic treatment. Additionally, most of the biodistribution and exposure studies for GTx and CTx are carried out in nonclinical studies since imitations of the samples collected in clinic. So, nonclinical studies will be used in predicting dosing in humans and exposure.

PK measurements in cell therapies might be achieved by enumerating the intended cells when proliferation of these cells is expected. The expression of the transgene (PD) may be easily measured by RT-PCR or by detection of expressed protein by traditional technologies such as LBA or other platforms.

Overall, in GTx and CTx, most of the technologies include traditional BA techniques, but not in the same importance. Typically, the focus is in NAb, vector copy number (VCN), and biomarkers. PK is limited to cell therapies and oligonucleotide-based therapies. The biological responses to gene products can be complex and may include cytokine release syndrome and interference with various biomarker measurements, demanding long-term oversight. Bioanalysis for emerging modalities is immensely complex; they demand considerations above and beyond what is typical for traditional bioanalysis and face major challenges particularly navigating through regulatory requirements, data reporting, and managing logistical demands of associated biospecimen handling.

Disclaimer Any opinions or forward-looking statements expressed are those of the authors and may not reflect views held by their employers (Moderna Inc. for Darshana Jani, Wave Life Sciences for Ramakrishna Boyanapalli, and Sangamo Therapeutics for Liching Cao).

References

Agency EM. Guideline on immunogenicity assessment of therapeutic proteins, Committee for Medicinal Products for Human Use (CHMP) London, UK 2017.

Awasthi R, Pacaud L, Waldron E, Tam CS, Jager U, Borchmann P, Jaglowski S, Foley SR, van Besien K, Wagner-Johnston ND, Kersten MJ, Schuster SJ, Salles G, Maziarz RT, Anak O, Del Corral C, Chu J, Gershgorin I, Pruteanu-Malinici I, Chakraborty A, Mueller KT, Waller EK. Tisagenlecleucel cellular kinetics, dose, and immunogenicity in relation to clinical factors in relapsed/refractory DLBCL. Blood Adv. 2020;4(3):560–72.

BioMarin. ISTH global seroprevalance presentation. 2019.

Bourdage JS, Cook CA, Farrington DL, Chain JS, Konrad RJ. An affinity capture elution (ACE) assay for detection of anti-drug antibody to monoclonal antibody therapeutics in the presence of high levels of drug. J Immunol Methods. 2007;327(1–2):10–7.

Breton C, Clark PM, Wang L, Greig JA, Wilson JM. ITR-Seq, a next-generation sequencing assay, identifies genome-wide DNA editing sites in vivo following adeno-associated viral vector-mediated genome editing. BMC Genomics. 2020;21(1):239.

Britten CM, Walter S, Janetzki S. Immunological monitoring to rationally guide AAV gene therapy. Front Immunol. 2013;4:273.

Buttel IC, Chamberlain P, Chowers Y, Ehmann F, Greinacher A, Jefferis R, Kramer D, Kropshofer H, Lloyd P, Lubiniecki A, Krause R, Mire-Sluis A, Platts-Mills T, Ragheb JA, Reipert BM, Schellekens H, Seitz R, Stas P, Subramanyam M, Thorpe R, Trouvin JH, Weise M, Windisch J, Schneider CK. Taking immunogenicity assessment of therapeutic proteins to the next level. Biologicals. 2011;39(2):100–9.

Charlesworth CT, Deshpande PS, Dever DP, Camarena J, Lemgart VT, Cromer MK, Vakulskas CA, Collingwood MA, Zhang L, Bode NM, Behlke MA, Dejene B, Cieniewicz B, Romano R, Lesch BJ, Gomez-Ospina N, Mantri S, Pavel-Dinu M, Weinberg KI, Porteus MH. Identification of preexisting adaptive immunity to Cas9 proteins in humans. Nat Med. 2019;25(2):249–54.

Chen F, Fraietta JA, June CH, Xu Z, Joseph Melenhorst J, Lacey SF. Engineered T cell therapies from a drug development viewpoint. Engineering. 2019;5(1):140–9.

Cheung R, deHart GW, Jesaitis L, Zoog SJ, Melton AC. Detection of antibodies that neutralize the cellular uptake of enzyme replacement therapies with a cell-based assay. J Vis Exp. 2018;139:57777.

Colella P, Ronzitti G, Mingozzi F. Emerging issues in AAV-mediated in vivo gene therapy. Mol Ther Methods Clin Dev. 2018;8:87–104.

Concolino D, Deodato F, Parini R. Enzyme replacement therapy: efficacy and limitations. Ital J Pediatr. 2018;44(Suppl 2):120.

Demchuk MP, Klunnyk IO, Matiyashchuk IM, Sych N, Sinelnyk A, Novytska A, Sorochynska K. Efficacy of fetal stem cells use in complex treatment of patients with insulin-resistant type 2 diabetes mellitus. J Stem Cell Res Ther. 2016;6(6):342.

Doshi BS, Arruda VR. Gene therapy for hemophilia: what does the future hold? Ther Adv Hematol. 2018;9(9):273–93.

Dovgan I, Koniev O, Kolodych S, Wagner A. Antibody-oligonucleotide conjugates as therapeutic, imaging, and detection agents. Bioconjug Chem. 2019;30(10):2483–501.

Ertl HCJ, High KA. Impact of AAV capsid-specific T-cell responses on design and outcome of clinical gene transfer trials with recombinant adeno-associated viral vectors: an evolving controversy. Hum Gene Ther. 2017;28(4):328–37.

Falese L, Sandza K, Yates B, Triffault S, Gangar S, Long B, Tsuruda L, Carter B, Vettermann C, Zoog SJ, Fong S. Strategy to detect pre-existing immunity to AAV gene therapy. Gene Ther. 2017;24(12):768–78.

FDA, U. Guidance for industry: guidance for human somatic cell therapy and gene therapy. US Department of Health and Human Services, Rockville, MD, USA; 1998 **1**: 30.

FDA, U. Recommendations for microbial vectors used for gene therapy—guidance for industry. Silver Spring, MD: Food and Drug Administration; 2016.

FDA, U. Bioanalytical method validation guidance for industry, US Department of Health and Human Services. Food and Drug Administration, Center for Drug Evaluation and Research (CDER), Center for Veterinary Medicine (CVM), Biopharmaceutics: 2018; 1–44.

FDA, U. Draft guidance for industry: fabry disease: developing drugs for treatment. U.S. Department of Health and Human Services, Food and Drug Administration (FDA), Center for Biologics Evaluation and Research (CBER); 2019a.

FDA, U. Guidance for industry, immunogenicity testing of therapeutic protein products–developing and validating assays for anti-drug antibody detection; 2019b.

FDA, U. Guidance for industry: human gene therapy for hemophilia. U.S. Department of Health and Human Services, Food and Drug Administration (FDA), Center for Biologics Evaluation and Research (CBER) 2020a.

FDA, U. Guidance for industry: human gene therapy for rare disease. U.S. Department of Health and Human Services, Food and Drug Administration (FDA), Center for Biologics Evaluation and Research (CBER); 2020b.

FDA, U. Guidance for industry: human gene therapy for retinal disorders. U.S. Department of Health and Human Services, Food and Drug Administration (FDA), Center for Biologics Evaluation and Research (CBER); 2020c.

Goh JB, Ng SK. Impact of host cell line choice on glycan profile. Crit Rev Biotechnol. 2018;38(6):851–67.

Gomez-Aguado I, Rodriguez-Castejon J, Vicente-Pascual M, Rodriguez-Gascon A, Solinis MA, Del Pozo-Rodriguez A. Nanomedicines to deliver mRNA: state of the art and future perspectives. Nanomaterials (Basel). 2020;10:2.

Gorovits B, Koren E. Immunogenicity of chimeric antigen receptor T-cell therapeutics. BioDrugs. 2019;33(3):275–84.

Guo P, Zhang J, Chrzanowski M, Huang J, Chew H, Firrman JA, Sang N, Diao Y, Xiao W. Rapid AAV-neutralizing antibody determination with a cell-binding assay. Mol Ther Methods Clin Dev. 2019;13:40–6.

Gupta S, Richards S, Amaravadi L, Piccoli S, Desilva B, Pillutla R, Stevenson L, Mehta D, Carrasco-Triguero M, Neely R, Partridge M, Staack RF, Zhao X, Gorovits B, Kolaitis G, Sumner G, Stubenrauch KG, Zou L, Amur S, Beaver C, Berger I, Berisha F, Birnboeck H, Bower J, Cho SJ, Cludts I, Cocea L, Donato LD, Fischer S, Fraser S, Garofolo F, Haidar S, Haulenbeek J, Hottenstein C, Hu J, Ishii-Watabe A, Islam R, Jani D, Kadavil J, Kamerud J,

Kramer D, Kurki P, MacMannis S, McNally J, Mullan A, Papadimitriou A, Pedras-Vasconcelos J, Ray S, Safavi A, Saito Y, Savoie N, Fjording MS, Scheibner K, Smeraglia J, Song A, Stouffer B, Tampal N, der Strate BV, Verch T, Welink J, Xu Y, Yang TY, Yengi L, Zeng J, Zhang Y, Zhang Y, Zoog S. 2017 White paper on recent issues in bioanalysis: a global perspective on immunogenicity guidelines & biomarker assay performance (part 3—LBA: immunogenicity, biomarkers and PK assays). Bioanalysis. 2017;9(24):1967–96.

High KA, Roncarolo MG. Gene therapy. N Engl J Med. 2019;381(5):455–64.

Hu Y, Huang J. The chimeric antigen receptor detection toolkit. Front Immunol. 2020;11:1770.

Janetzki S, Panageas KS, Ben-Porat L, Boyer J, Britten CM, Clay TM, Kalos M, Maecker HT, Romero P, Yuan J, Kast WM, Hoos A, Elispot Proficiency Panel of the C. V. C. I. A. W. G. Results and harmonization guidelines from two large-scale international Elispot proficiency panels conducted by the cancer vaccine consortium (CVC/SVI). Cancer Immunol Immunother. 2008;57(3):303–15.

Jani D, Marsden R, Mikulskis A, Gleason C, Klem T, Krinos Fiorotti C, Myler H, Yang L, Fiscella M. Recommendations for the development and validation of confirmatory anti-drug antibody assays. Bioanalysis. 2015;7(13):1619–31.

Kaiser J. How safe is a popular gene therapy vector? Science. 2020;367(6474):131.

Kakkanaiah VN, Lang KR, Bennett PK. Flow cytometry in cell-based pharmacokinetics or cellular kinetics in adoptive cell therapy. Bioanalysis. 2018;10(18):1457–9.

Kowalski PS, Rudra A, Miao L, Anderson DG. Delivering the messenger: advances in technologies for therapeutic mRNA delivery. Mol Ther. 2019;27(4):710–28.

Kramer MG, Masner M, Ferreira FA, Hoffman RM. Bacterial therapy of cancer: promises, limitations, and insights for future directions. Front Microbiol. 2018;9:16.

Kruzik A, Koppensteiner H, Fetahagic D, Hartlieb B, Dorn S, Romeder-Finger S, Coulibaly S, Weber A, Hoellriegl W, Horling FM, Scheiflinger F, Reipert BM, de la Rosa M. Detection of biologically relevant low-Titer neutralizing antibodies against adeno-associated virus require sensitive in vitro assays. Hum Gene Ther Methods. 2019;30(2):35–43.

Kubiak RJ, Lee N, Zhu Y, Franch WR, Levitskaya SV, Krishnan SR, Abraham V, Akufongwe PF, Larkin CJ, White WI. Storage conditions of conjugated reagents can impact results of immunogenicity assays. J Immunol Res. 2016;2016:1485615.

Kulkarni JA, Cullis PR, van der Meel R. Lipid nanoparticles enabling gene therapies: from concepts to clinical utility. Nucleic Acid Ther. 2018;28(3):146–57.

Kuranda K, Jean-Alphonse P, Leborgne C, Hardet R, Collaud F, Marmier S, Costa Verdera H, Ronzitti G, Veron P, Mingozzi F. Exposure to wild-type AAV drives distinct capsid immunity profiles in humans. J Clin Invest. 2018;128(12):5267–79.

Leipold MD, Newell EW, Maecker HT. Multiparameter phenotyping of human PBMCs using mass cytometry. Methods Mol Biol. 2015;1343:81–95.

Li Z, Hayman RB, Walt DR. Detection of single-molecule DNA hybridization using enzymatic amplification in an array of femtoliter-sized reaction vessels. J Am Chem Soc. 2008;130(38):12622–3.

Long BR, Sandza K, Holcomb J, Crockett L, Hayes GM, Arens J, Fonck C, Tsuruda LS, Schweighardt B, O'Neill CA, Zoog S, Vettermann C. The impact of pre-existing immunity on the non-clinical pharmacodynamics of AAV5-based gene therapy. Mol Ther Methods Clin Dev. 2019;13:440–52.

Ma H, Bell KN, Loker RN. qPCR and qRT-PCR analysis: regulatory points to consider when conducting biodistribution and vector shedding studies. Mol Ther Methods Clin Dev. 2021;20:152–68.

Ma M, Balasubramanian N, Dodge R, Zhang Y. Challenges and opportunities in bioanalytical support for gene therapy medicinal product development. Bioanalysis. 2017;9(18):1423–30.

Majowicz A, Nijmeijer B, Lampen MH, Spronck L, de Haan M, Petry H, van Deventer SJ, Meyer C, Tangelder M, Ferreira V. Therapeutic hFIX activity achieved after single AAV5-hFIX treatment in Hemophilia B patients and NHPs with pre-existing anti-AAV5 NABs. Mol Ther Methods Clin Dev. 2019;14:27–36.

Manno CS, Pierce GF, Arruda VR, Glader B, Ragni M, Rasko JJ, Ozelo MC, Hoots K, Blatt P, Konkle B, Dake M, Kaye R, Razavi M, Zajko A, Zehnder J, Rustagi PK, Nakai H, Chew A, Leonard D, Wright JF, Lessard RR, Sommer JM, Tigges M, Sabatino D, Luk A, Jiang H, Mingozzi F, Couto L, Ertl HC, High KA, Kay MA. Successful transduction of liver in hemophilia by AAV-factor IX and limitations imposed by the host immune response. Nat Med. 2006;12(3):342–7.

Masat E, Pavani G, Mingozzi F. Humoral immunity to AAV vectors in gene therapy: challenges and potential solutions. Discov Med. 2013;15(85):379–89.

Mauhin W, Lidove O, Amelin D, Lamari F, Caillaud C, Mingozzi F, Dzangue-Tchoupou G, Arouche-Delaperche L, Douillard C, Dussol B, Leguy-Seguin V, D'Halluin P, Noel E, Zenone T, Matignon M, Maillot F, Ly KH, Besson G, Willems M, Labombarda F, Masseau A, Lavigne C, Froissart R, Lacombe D, Ziza JM, Hachulla E, Benveniste O. Deep characterization of the anti-drug antibodies developed in Fabry disease patients, a prospective analysis from the French multicenter cohort FFABRY. Orphanet J Rare Dis. 2018;13(1):127.

Meliani A, Leborgne C, Triffault S, Jeanson-Leh L, Veron P, Mingozzi F. Determination of anti-adeno-associated virus vector neutralizing antibody titer with an in vitro reporter system. Hum Gene Ther Methods. 2015;26(2):45–53.

Melton AC, Soon RK Jr, Tompkins T, Long B, Schweighardt B, Qi Y, Vitelli C, Bagri A, Decker C, O'Neill CA, Zoog SJ, Jesaitis L. Antibodies that neutralize cellular uptake of elosulfase alfa are not associated with reduced efficacy or pharmacodynamic effect in individuals with Morquio a syndrome. J Immunol Methods. 2017;440:41–51.

Mfarrej B, Gaude J, Couquiaud J, Calmels B, Chabannon C, Lemarie C. Validation of a flow cytometry-based method to quantify viable lymphocyte subtypes in fresh and cryopreserved hematopoietic cellular products. Cytotherapy. 2021;23(1):77–87.

Millipore E. Elispot assays: state-of-the-art tools for functional analysis of cellular immunology. White paper; n.d.

Milone MC, Bhoj VG. The pharmacology of T cell therapies. Mol Ther Methods Clin Dev. 2018;8:210–21.

Mingozzi F, High KA. Immune responses to AAV vectors: overcoming barriers to successful gene therapy. Blood. 2013;122(1):23–36.

Mittag A, Tarnok A. Recent advances in cytometry applications: preclinical, clinical, and cell biology. Methods Cell Biol. 2011;103:1–20.

Mount NM, Ward SJ, Kefalas P, Hyllner J. Cell-based therapy technology classifications and translational challenges. Philos Trans R Soc Lond Ser B Biol Sci. 2015;370(1680):20150017.

Mueller KT, Waldron E, Grupp SA, Levine JE, Laetsch TW, Pulsipher MA, Boyer MW, August KJ, Hamilton J, Awasthi R, Stein AM, Sickert D, Chakraborty A, Levine BL, June CH, Tomassian L, Shah SS, Leung M, Taran T, Wood PA, Maude SL. Clinical pharmacology of tisagenlecleucel in B-cell acute lymphoblastic leukemia. Clin Cancer Res. 2018;24(24):6175–84.

Nathwani AC, Rosales C, McIntosh J, Rastegarlari G, Nathwani D, Raj D, Nawathe S, Waddington SN, Bronson R, Jackson S, Donahue RE, High KA, Mingozzi F, Ng CY, Zhou J, Spence Y, McCarville MB, Valentine M, Allay J, Coleman J, Sleep S, Gray JT, Nienhuis AW, Davidoff AM. Long-term safety and efficacy following systemic administration of a self-complementary AAV vector encoding human FIX pseudotyped with serotype 5 and 8 capsid proteins. Mol Ther. 2011a;19(5):876–85.

Nathwani AC, Tuddenham EG, Rangarajan S, Rosales C, McIntosh J, Linch DC, Chowdary P, Riddell A, Pie AJ, Harrington C, O'Beirne J, Smith K, Pasi J, Glader B, Rustagi P, Ng CY, Kay MA, Zhou J, Spence Y, Morton CL, Allay J, Coleman J, Sleep S, Cunningham JM, Srivastava D, Basner-Tschakarjan E, Mingozzi F, High KA, Gray JT, Reiss UM, Nienhuis AW, Davidoff AM. Adenovirus-associated virus vector-mediated gene transfer in hemophilia B. N Engl J Med. 2011b;365(25):2357–65.

Palffy R, Gardlik R, Hodosy J, Behuliak M, Resko P, Radvansky J, Celec P. Bacteria in gene therapy: bactofection versus alternative gene therapy. Gene Ther. 2006;13(2):101–5.

Pan L, Hilton K, Nadeau M, McCauley T, Qiu Y. A comparison study of bioanalytical methods for characterization of anti-idursulfase antibodies. Bioanalysis. 2017;9(16):1237–46.

Partridge MA, Purushothama S, Elango C, Lu Y. Emerging technologies and generic assays for the detection of anti-drug antibodies. J Immunol Res. 2016;2016:6262383.

Piccoli S, Mehta D, Vitaliti A, Allinson J, Amur S, Eck S, Green C, Hedrick M, Hopper S, Ji A, Joyce A, Litwin V, Maher K, Mathews J, Peng K, Safavi A, Wang YM, Zhang Y, Amaravadi L, Palackal N, Thankamony S, Beaver C, Bame E, Emrich T, Grimaldi C, Haulenbeek J, Joyce A, Kakkanaiah V, Lanham D, Maher K, Mayer A, Trampont PC, Vermet L, Dakappagari N, Fleener C, Garofolo F, Rogers C, Tangri S, Xu Y, Liang M, Rajadhyaksha M, Richards S, Schweighardt B, Purushothama S, Baltrukonis D, Brumm J, Cherry E, Delcarpini J, Gleason C, Kirshner S, Kubiak R, Pan L, Partridge M, Pedras-Vasconcelos J, Qu Q, Skibeli V, Saunders TS, Staack RF, Stubenrauch K, Torri A, Verthelyi D, Yan H, Gorovits B, Palmer R, Milton M, Long B, Corsaro B, Farrokhi V, Fiscella M, Henderson N, Jawa V, McNally J, Murphy R, Waldner H, Yang TY. 2019 White paper on recent issues in bioanalysis: FDA immunogenicity guidance, gene therapy, critical reagents, biomarkers and flow cytometry validation (part 3 - recommendations on 2019 FDA immunogenicity guidance, gene therapy bioanalytical challenges, strategies for critical reagent management, biomarker assay validation, flow cytometry validation & CLSI H62). Bioanalysis. 2019;11(24):2207–44.

Pien GC, Basner-Tschakarjan E, Hui DJ, Mentlik AN, Finn JD, Hasbrouck NC, Zhou S, Murphy SL, Maus MV, Mingozzi F, Orange JS, High KA. Capsid antigen presentation flags human hepatocytes for destruction after transduction by adeno-associated viral vectors. J Clin Invest. 2009;119(6):1688–95.

Pol E, Roos H, Markey F, Elwinger F, Shaw A, Karlsson R. Evaluation of calibration-free concentration analysis provided by Biacore systems. Anal Biochem. 2016;510:88–97.

Roidos P, Sungalee S, Benfatto S, Sercin O, Stutz AM, Abdollahi A, Mauer J, Zenke FT, Korbel JO, Mardin BR. A scalable CRISPR/Cas9-based fluorescent reporter assay to study DNA double-strand break repair choice. Nat Commun. 2020;11(1):4077.

Samulski RJ, Muzyczka N. AAV-mediated gene therapy for research and therapeutic purposes. Annu Rev Virol. 2014;1(1):427–51.

Santeramo I, Bagnati M, Harvey EJ, Hassan E, Surmacz-Cordle B, Marshall D, Di Cerbo V. Vector copy distribution at a single-cell level enhances analytical characterization of gene-modified cell therapies. Mol Ther Methods Clin Dev. 2020;17:944–56.

Santiago-Toledo G, Georgiou M, Dos Reis J, Roberton VH, Valinhas A, Wood RC, Phillips JB, Mason C, Li D, Li Y, Sinden JD, Choi D, Jat PS, Wall IB. Generation of c-MycER(TAM)-transduced human late-adherent olfactory mucosa cells for potential regenerative applications. Sci Rep. 2019;9(1):13190.

Savva M. Fundamental concepts of dosage calculations. Pharmaceutical calculations: a conceptual approach. Cham: Springer International Publishing; 2019. p. 301–27.

Scallan CD, Jiang H, Liu T, Patarroyo-White S, Sommer JM, Zhou S, Couto LB, Pierce GF. Human immunoglobulin inhibits liver transduction by AAV vectors at low AAV2 neutralizing titers in SCID mice. Blood. 2006;107(5):1810–7.

Shankar G, Devanarayan V, Amaravadi L, Barrett YC, Bowsher R, Finco-Kent D, Fiscella M, Gorovits B, Kirschner S, Moxness M, Parish T, Quarmby V, Smith H, Smith W, Zuckerman LA, Koren E. Recommendations for the validation of immunoassays used for detection of host antibodies against biotechnology products. J Pharm Biomed Anal. 2008;48(5):1267–81.

Shirley JL, de Jong YP, Terhorst C, Herzog RW. Immune responses to viral gene therapy vectors. Mol Ther. 2020;28(3):709–22.

Simhadri VL, McGill J, McMahon S, Wang J, Jiang H, Sauna ZE. Prevalence of pre-existing antibodies to CRISPR-associated nuclease Cas9 in the USA population. Mol Ther Methods Clin Dev. 2018;10:105–12.

Smith HW, Butterfield A, Sun D. Detection of antibodies against therapeutic proteins in the presence of residual therapeutic protein using a solid-phase extraction with acid dissociation (SPEAD) sample treatment prior to ELISA. Regul Toxicol Pharmacol. 2007;49(3):230–7.

Solomon M, Muro S. Lysosomal enzyme replacement therapies: historical development, clinical outcomes, and future perspectives. Adv Drug Deliv Rev. 2017;118:109–34.

Steinmetz KL, Spack EG. The basics of preclinical drug development for neurodegenerative disease indications. BMC Neurol. 2009;9(Suppl 1):S2.

Stevenson L, Richards S, Pillutla R, Torri A, Kamerud J, Mehta D, Keller S, Purushothama S, Gorovits B, Litwin V, Stebbins C, Marini J, Beaver C, Sperinde G, Siguenza P, Staack RF, Qiu Y, Amaravadi L, Amur S, Fleener CA, Baltrukonis D, Catlett I, Cherry E, Chung S, Cludts I, Donato LD, Fischer S, Fraser S, Garofolo F, Green C, Gunn G, Haidar S, Haulenbeek J, Henderson N, Hopper S, Ishii-Watabe A, Islam R, Janelsins B, Jawa V, Kakkanaiah V, Kamondi S, Kolaitis G, Kubiak RJ, Kumar S, Kurki P, Liang M, Liu P, Maxfield K, Myler H, Palackal N, Palmer R, Pedras-Vasconcelos J, Piccoli S, Rhyne P, Saito Y, Savoie N, Schick E, Schweighardt B, Shih J, Song A, Sriraman P, Staelens L, Sumner G, Sun Y, Ullmann M, Verthelyi D, Wadhwa M, Wang YM, Xu Y, Yan H, Yang TY, Zeng R. 2018 White paper on recent issues in bioanalysis: focus on flow cytometry, gene therapy, cut points and key clarifications on BAV (part 3—LBA/cell-based assays: immunogenicity, biomarkers and PK assays). Bioanalysis. 2018;10(24):1973–2001.

Tary-Lehmann M, Hamm CD and Lehmann PV. Validating reference samples for comparison in a regulated ELISPOT assay. Validation of cell-based assays in the GLP setting: 2008; 127–146.

Torchinsky D, Ebenstein Y. Sizing femtogram amounts of dsDNA by single-molecule counting. Nucleic Acids Res. 2016;44(2):e17.

Welink J, Xu Y, Yang E, Wilson A, Henderson N, Luo L, Fraser S, Kavita U, Musuku A, James C, Fraier D, Zhang Y, Goykhman D, Summerfield S, Woolf E, Verhaeghe T, Hughes N, Behling A, Brown K, Bulychev A, Buonarati M, Cherry E, Cho SJ, Cludts I, Dillen L, Dodge R, Edmison A, Garofolo F, Green R, Haidar S, Hottenstein C, Ishii-Watabe A, Jang HG, Ji A, Jones B, Kassim S, Ma M, Lima Santos GM, Norris DA, Owen T, Piccoli S, Ramanathan R, Rohl I, Rosenbaum AI, Saito Y, Sangster T, Savoie N, Stebbins C, Sydor J, de Merbel NV, Verthelyi D, Vinter S, Whale E. 2018 White paper on recent issues in bioanalysis: 'A global bioanalytical community perspective on last decade of incurred samples reanalysis (ISR)' (part 1—small molecule regulated bioanalysis, small molecule biomarkers, peptides & oligonucleotide bioanalysis). Bioanalysis. 2018;10(22):1781–801.

Weltner J, Anisimov A, Alitalo K, Otonkoski T, Trokovic R. Induced pluripotent stem cell clones reprogrammed via recombinant adeno-associated virus-mediated transduction contain integrated vector sequences. J Virol. 2012;86(8):4463–7.

Yamamoto S, Matsumoto SI, Shimizu H, Hirabayashi H. Quantitative application of flow cytometry for the analysis of circulating human T cells: a preclinical pharmacokinetic study. Drug Metab Pharmacokinet. 2020;35(2):207–13.

Yang TY, Doddareddy R. Considerations in the development and validation of real-time quantitative polymerase chain reaction and its application in regulated bioanalysis to characterize the cellular kinetics of CAR-T products in clinical studies. Bioanalysis. 2020;13:115.

Chapter 6
Peptides and Oligonucleotide-Based Therapy: Bioanalytical Challenges and Practical Solutions

Ramakrishna Boyanapalli, Inderpal Singh, and Morse Faria

Abstract Peptide and Oligonucleotide (ON) are gaining popularity with multiple regulatory approvals in recent years. Most of the approvals are in genetic disorders for various disease areas. Peptides and ON molecules are considered as small-molecule drugs that are synthesized with various chemical and structural modifications to improve their physicochemical and biological properties, such as efficacy, stability, bioavailability, safety, etc. These modifications require thorough understanding of these molecules and exclusive strategies to quantify these molecules in various biological matrices. This chapter provides an overview of currently available methodologies and strategies employed for bioanalytical and immunogenicity assessment of peptides and ON therapeutic modalities.

Keywords Oligonucleotides · Peptides · LC-MS/MS · Ligand binding assays · qPCR · Pharmacokinetics · Immunogenicity

6.1 Introduction

New therapeutic modalities such as short chains of amino acids or nucleic acids called peptide or oligonucleotide (ON) therapeutics, respectively, are becoming prominent therapies with multiple drug receiving regulatory approvals in the recent years. A comprehensive list of peptide drugs (de la Torre and Albericio 2020) and ON drugs (Roberts et al. 2020) is provided in these recent review articles. The list is anticipated to grow since a plethora of biotech and pharmaceutical companies are

R. Boyanapalli (✉)
Bioanalytical and biomarker Development, Wave Life Sciences, Cambridge, MA, USA

I. Singh
Spark Therapeutics, Philadelphia, PA, USA

M. Faria
Clinical Research Group, Thermo Fisher Scientific, Richmond, VA, USA

S. Kumar (ed.), *An Introduction to Bioanalysis of Biopharmaceuticals*, AAPS Advances in the Pharmaceutical Sciences Series 57,
https://doi.org/10.1007/978-3-030-97193-9_6

continuing their efforts in discovery and development of these modalities. Over time, these therapeutics have evolved chemically and structurally to overcome challenges, such as stability, bioavailability, tolerability, drug delivery, therapeutic coverage, etc. The chemical and structural changes have substantially modified these modalities from their naturally occurring molecular counterparts. These modifications are made mostly to the molecules' backbone chemistry, amino acid or nucleotide types, etc. With these changes, the molecules can be administered nascently in formulation buffer (unconjugated / gymnotic) or conjugated to lipid or glycan moieties to further enhance drug bioavailability (Lau and Dunn 2018; Lee et al. 2019; Roberts et al. 2020). In addition, a new class of modalities, known as peptide nucleic acids (PNAs), which are the combination of peptides and ON are also developed. The PNAs are shown to increase specificity and sequence selectivity towards their target DNA or RNA (Quijano et al. 2017; Wu et al. 2017; Montazersaheb et al. 2018). The bioanalysis of PNA molecules is out of the scope of this chapter.

Even though the peptide and ON therapeutic modalities have natural biological origins, because of their molecular size and feasibility of manufacturing in biological free systems, these modalities are considered as small-molecule drugs (Xu et al. 2019). However, compared to other small-molecule drugs, these therapeutics are relatively large in size and complex. So, bioanalysis of these therapeutic modalities can be challenging.

6.1.1 Peptide Drugs and Modifications

Since the approval of insulin, peptides therapeutics have become an important class of therapeutic agents. More than 80 peptide drugs have been approved in the USA and other major markets for a wide range of diseases, including diabetes, cancer, HIV infection, multiple sclerosis, osteoporosis, and chronic pain (Muttenthaler et al. 2021). Peptides as a therapeutic modality have several desirable attributes such as low toxicity, high specificity, and high potency. However, development of therapeutic peptides also has several hurdles including challenging and costly synthesis, need for parenteral administration due to low oral bioavailability, solubility issues, high clearance rates, and high immunogenic risk. Peptide drugs can be synthetically manufactured or extracted in polypeptide forms from natural sources to mimic endogenous hormones, neurotransmitters, growth factors, signaling molecules, etc. Natural polypeptides have lower stability. So, these peptides will be dosed frequently to maintain required efficacy (Lee et al. 2019). To overcome these challenges, amino acids in a peptide drug can be modified to improve their characteristics. For example, icantibant, a bradykinin antagonist, has D-phenylalanine in place of a proline at the seventh position (Charignon et al. 2012). Other varieties of synthetic amino acids were designed to enhance resistance to proteases. These modifications can be stereo-isomers, β3 analogues of arginine, homoarginine, lysine or ornithine. In addition, aromatic amino acids with replaced β-methyl group were also synthesized to increase stability without compromising the drug function (Lee et al. 2019).

While designing the changes, the immunogenic nature of these peptides will be monitored since non-native structures can be immunogenic. Encouragingly, these modifications can assist in bioanalysis due to their peculiar structures. That advantage might be missing in natural polypeptide drug bioanalysis due to interference from any endogenous form of the peptide.

6.1.2 Oligonucleotide Drugs and Modifications

Majority of these mechanisms of actions shown by ON therapeutics are target-specific gene silencing using small interference RNA (siRNA), microRNA (miRNA), and antisense oligonucleotides (ASOs); RNA editing using adenosine deaminase acting on RNA (ADAR); RNA splicing modulation; short single-stranded (SS) oligonucleotides that can selectively bind ligands or protein targets (Aptamers); etc. Based on the mode of action, ON drugs can be single-stranded (SS) or double-stranded (DS) nucleic acid fragments (Thiel and Giangrande 2009; Merkle et al. 2019; Bajan and Hutvagner 2020; Roberts et al. 2020). Some of the examples of ON chemistry and backbone modifications are phosphorothioate (PS) replacing phosphodiester in the backbone linkages; 2′ O-methoxyethyl (2′-MOE), 2′ O-methyl (2′-MO), and 2′-fluoro (2′-F) for sugar modifications; locked nucleic acids (LNA), ethyl bridged nucleic acids (ENA), and constrained ethyl bridged nucleic acids (cET) as base modifications or bridged nucleotides; and phosphorodiamidate morpholino ON (PMO; peptide nucleic acids (PNA) and tricyclo DNA (tcDNA) as other chemistry modifications. A few modifications, such as the phosphodiester backbone, can form a chiral site, which results in *Rp-Sp* stereoisomers. Using the chiral site stereochemistry, stereo-pure ONs are also being synthesized and tested as drugs (Iwamoto et al. 2017; Roberts et al. 2020; Featherston et al. 2021; Liczner et al. 2021; Liu et al. 2021). The stereo-pure ON chemistry is evolving with different modifications. Bioanalysis methods for quantifying stereo-pure ON may not require any specific tools. Specific methods such as ion mobility can be used to separate diastereomers in a mixture of ON isomers (Sutton et al. 2021). Certain ON are also conjugated to a polymer such as polyethylene glycol (PEG) to increase bioavailability (Shokrzadeh et al. 2014). Similar to peptides, these modifications can result in drug-dependent immune response. Overall, ONs are known to show innate response and humoral (adaptive) response with moderate adverse effects (Bodera et al. 2012). During drug discovery stages, modifications were selected with minimal immune response. Depending on the modifications, the bioanalysis can be supportive or challenging.

The consequences of an immune reaction to therapeutic molecules range from transient appearance without any clinical significance to life-threatening conditions. Despite the relatively small size of these therapeutic modalities, it has been recognized that these molecules may induce unwanted humoral and cellular immune responses. Due to various modifications to the nucleic acids and amino acids to increase stability, immune response to these molecules can be higher than expected

(Wang et al. 2015). Many factors may influence the immunogenicity of therapeutic molecules. Both patient-related and product-related factors may affect immunogenicity of therapeutic protein products. Patient-related factors that might predispose to an immune response include underlying disease, genetic background, and immune status, including immunomodulating therapy. Product-related factors also influence the likelihood of an immune response, such as intensity of treatment (route of administration), source of the molecules, manufacturing process (impurity profile, contaminants), formulation, and stability characteristics (degradation products, aggregates) of a given molecule and dose (dosing interval and duration of treatment). It is essential to consider adopting a risk-based strategy by development of adequate assays for evaluating and predicting immune responses to these new therapeutic modalities. Immune response to ON is noticed to be subtle; it might be predominantly innate and cellular response compared to humoral response (Hornung et al. 2005; Judge et al. 2005). However, checking for anti-drug antibodies (ADAs) is required by the agency, and testing for neutralizing antibodies for ON therapeutics may be uncertain or irrelevant due to the intracellular mechanism of action. The immunogenicity assessment strategies for these new modalities are getting traction to obtain additional regulations in place.

In this chapter, he bioanalytical strategies for pharmacokinetic (PK) and immunogenicity evaluations of peptides and ON therapeutic modalities are presented. Similar to small- and large-molecule therapeutic molecules, bioanalysis plays an integral role in understanding the disposition (PK, biodistribution, metabolism and immunogenicity) of peptides and ON therapeutic modalities (Kay and Roberts 2012; Ewles et al. 2014; Wang and Ji 2016; Lorenson et al. 2019). The later part of the chapter presents various analytical platforms and assay development approaches that can be used for the quantitation of peptides and ON modalities. The chapter ends with a section on regulatory requirements and current industry practices in bioanalysis of these therapeutic modalities (FDA 2021a, b, c).

6.2 Peptides Bioanalysis for Pharmacokinetic Evaluations

According to the definition by US FDA, a peptide is any alpha amino acid polymer with a specific-defined sequence that is 40 amino acids or less in size (FDA 2018a, b). As a general rule, peptides are considered to have a mass less than 10 kDa.

Historically, bioanalysis of peptides for PK evaluations was carried out by ligand binding assays (LBAs) or immunoassays such as enzyme-linked immunosorbent assay (ELISA) and radioimmunoassay. Immunoassays, though sensitive and have high throughput, suffer from selectivity issues due to the inability of the capture antibodies to distinguish between structural similar peptides. Over the last two decades, liquid chromatography coupled tandem mass spectrometry (LC-MS/MS) or liquid chromatography coupled high-resolution mass spectrometry (LC-HRMS) has become the preferred bioanalytical platform for peptide PK bioanalysis, primarily due to high specificity of these methods and ease of method development (van den Broek et al. 2008; Rauh 2012; Chappell et al. 2014; Kay et al. 2016; Kang et al. 2020). For larger peptides, a hybrid assay format combining immunoaffinity enrichment and LC-MS/MS analysis has also been employed (Xu et al. 2010; Thomas et al. 2017; Xu et al. 2017, 2018). Other detectors such as ultraviolet (UV), florescence, or electrochemical detection, after liquid chromatography (LC) separation, are also employed for peptide bioanalysis. However, these detectors have limited sensitivity and lower specificity in comparison to mass spectrometry. A review paper published by Chappell et al. (2014). summarizes several qualified or validated bioanalytical methods for measurement of peptides for supporting drug development from discovery to clinical development (Chappell et al. 2014). In this chapter, we will focus on assay development on the LC-MS/MS platform.

6.2.1 Assay Development on LC-MS/MS Platform

For smaller peptides, i.e., less than 7 kDa, the common approach is to measure the intact peptide using LC-MS/MS. However, for larger peptides, there is a choice of measuring the intact peptide or a surrogate peptide after proteolytic digestion. The large peptides have lower sensitivity due to charge distribution across multiple charge states during ionization for mass spectrometric detection. The smaller size of surrogate peptide after proteolytic digestion enables higher sensitivity and hence can be the favored choice for bioanalysis of larger peptides. However, the use of a surrogate peptide does not always result in increased sensitivity. In some instances, the proteolytic digestion can adversely impact signal-to-noise ratio due to higher background interferences resulting from the increased complexity of the digested sample (Bronsema et al. 2013). Choosing a type of surrogate peptide is important since the surrogate peptide chosen may not derive during in vivo metabolism or modifications. Figure 6.1 illustrates the workflow for peptide bioanalysis by LC-MS/MS. Typically, stable labeled isotopes (SIL) form of the analyte peptide is used as internal standards for peptide bioanalysis by LC-MS/MS.

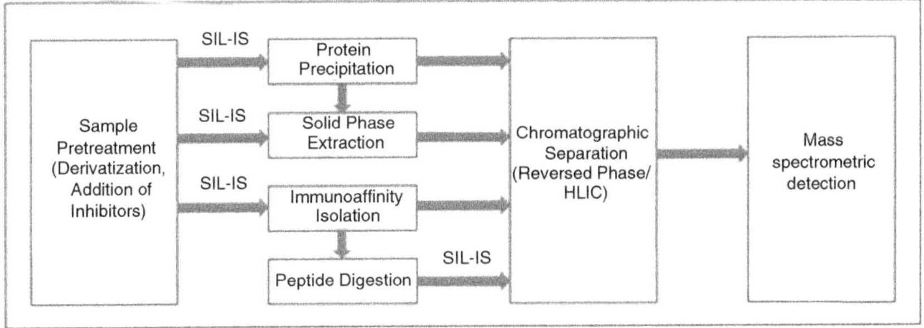

Fig. 6.1 Typical workflow of peptide bioanalysis by LC-MS/MS

6.2.2 Challenges in Sample Handling for Peptide Bioanalysis

6.2.2.1 Adsorption

Peptides are known to have non-specific binding to sample processing vessel surfaces at various steps of the bioanalytical method, i.e., from sample collection, storage, preparation of standard stock solutions, sample extraction, to even sample injection on the LC-MS/MS system (van Midwoud et al. 2007; Maes et al. 2014). Peptide adsorption can result in poor method performance by impacting linearity, sensitivity, accuracy and precision, and carryover. Hence, it is essential to evaluate possible losses due to non-specific binding during method development. The degree of adsorption is influenced by multiple factors such as structure, size, peptide amino acid composition, hydrophobicity, solvent composition, solvent pH, and the chemistry of the vessel surfaces (van den Broek et al. 2008; Yuan 2019). Positively charged peptides tend to adsorb with glass surfaces, while neutral peptides tend to adsorb to hydrophobic materials such as polypropylene. Hence, choosing containers with the suitable chemistry to minimize peptide non-specific binding is necessary. Adsorption will be more severe at low-concentration aqueous solutions. Addition of organic solvents such as acetonitrile, methanol, dimethyl sulfoxide, etc. in the stock dilution solvents helps improve peptide solubility and minimize non-specific binding. For peptides in biological matrices, the adsorption is significantly lower due to presence of abundant endogenous proteins that can preferentially bind to the adsorption sites. For biological matrices having lower endogenous proteins such as urine and cerebrospinal fluids, precautions need to be taken to avoid non-specific binding. For matrices such as urine and cerebrospinal fluids, it is advisable to evaluate non-specific binding during method development to minimize analyte losses during sample collection, processing, and storage. Matrix-free dilution during sample preparation should be avoided. If matrix-free dilution is required, bovine serum albumin or a surfactant such as Tween-20, Triton X-100, or CHAPS may be added to dilution solvent to minimize adsorption. If required, special containers having treated surface material, such as low-bind vials/plates, that minimizes peptide adsorption are used. Such containers are especially necessary for methods which use immunoaffinity enrichment during sample preparation.

6.2.2.2 Stability

Peptides are stable in solid state; however, in solution, they are susceptible to chemical degradation such as oxidation, reduction, deamidation, and hydrolysis (van den Broek et al. 2008, Yuan 2019). For example, peptides containing methionine are susceptible to oxidation, and peptides containing asparagine-glycine or aspartic acid-glycine are susceptible to isomerization and deamidation. When the peptide therapeutic has amino acids that are susceptible to degradations, precaution needs to be taken to ensure the peptide degradation is minimized during sample handling and extraction. Temperature control, pH control, and use of antioxidants have shown to minimize chemical degradation during bioanalysis (Li et al. 2011). Furthermore, peptides may undergo enzymatic cleavage due to proteases present in the biological matrix. Techniques used to minimize enzymatic degradation include extraction at lower temperature, i.e., on ice, addition of acids such as formic acid or hydrochloric acid, and addition of protease inhibitors or inhibitor cocktails may be added during sample collection (van den Broek et al. 2008; Yuan 2019). In a method, where the peptide was undergoing rapid enzymatic degradation, direct denaturation (collection of samples in organic solvent) was used to inactivate the proteases (Haynes et al. 2011).

6.2.3 Sample Preparation for Peptide Bioanalysis by LC-MS/MS

Protein precipitation (PPT), solid phase extraction (SPE), and immunoaffinity isolation are the three main techniques used for quantitation of peptide biotherapeutics by LC-MS. Liquid-liquid extraction (LLE), a popular sample preparation technique for small molecules, has limited application in peptide bioanalysis. This is primarily because most peptides have poor solubility in organic solvents due to their ionic nature. Hence, LLE is limited to non-ionic (hydrophobic) peptides such as cyclic peptides (Yuan 2019). Protein precipitation as a sample preparation technique is well established for small molecule bioanalysis. Protein precipitation (PPT) can be achieved using water-miscible organic solvents (e.g., acetonitrile, methanol, ethanol), acids (e.g., trichloroacetic acid), and salts (e.g., zinc sulphate, ammonium sulphate) (Polson et al. 2003). While using protein precipitation in peptide bioanalysis, it is necessary to ensure the peptide is not precipitating out. Large peptides are susceptible for co-precipitation due to lower aqueous solubility. In such instances, lower organic solvent volumes (e.g., a 2:1 volume ratio of acetonitrile to matrix) can be used to improve recovery (Chang et al. 2005). PPT is simple, fast, and easy to develop. However, it is comparatively limited in its capability for sample clean-up. A major drawback is its inability to remove phospholipids and hence susceptibility to matrix effects. PPT can be used as a sample preparation technique for bioanalytical methods to support early drug studies or those that do not require low detection limits. PPT can be combined with other clean-up techniques such as SPE to achieve higher clean-up and target lower detection limits.

SPE is another sample clean-up technique that is used either alone or in combination with other techniques such as PPT in peptide bioanalysis. Endogenous proteins, phospholipids, and salts from biological matrices are effectively removed while retaining the peptides on SPE columns. Depending on the physicochemical properties of the analyte peptide, such as polarity, basicity, charge, etc., the SPE column sorbent chemistry, loading, wash, and elution conditions are selected. Compared to protein precipitation and LLE, SPE sample extracts are much cleaner and hence are the preferred method to target low detection limits or when higher selectivity is required.

Immunoaffinity isolation has also emerged as sample preparation technique for peptide bioanalysis using LC-MS, especially when adequate sensitivity or selectivity is not achieved using conventional methods, i.e., PPT, LLE, and SPE. Immunoaffinity isolation is more expensive and requires the availability of a suitable capture reagent. For example, Xu et al. were able to achieve a detection limit of 1.00 and 0.500 ng/mL using PPT and SPE, respectively, for quantification of insulin glargine in human plasma. However, the target detection limit of 0.100 ng/mL was only achieved using an immunoaffinity isolation method using anti-insulin antibody as a capture reagent (Xu et al. 2017). If the analyte peptide is larger in size, the peptide can be digested to form smaller peptides and a surrogate peptide obtained after proteolysis is used for LC-MS/MS quantitation similar to hybrid LC-MS/MS methodology described in Chap. 2. The methodology for peptide bioanalysis using immunoaffinity coupled LC-MS/MS is well explored for measurement of endogenous peptides (Thomas et al. 2012).

6.2.4 Instrumentation for Peptide Bioanalysis by LC-MS/MS

Reversed-phase chromatography is the primary choice of chromatographic separation of peptide bioanalysis by LC-MS/MS. The chromatographic stationary phase is selected based on the physicochemical properties (mainly hydrophobicity) of the peptides. Retention of highly polar peptides is challenging using reversed-phase chromatography and may require the use of hydrophilic interaction chromatography (HILIC) for efficient separation. To achieve additional cleanup, many LC-MS/MS methods use a trap column prior analytical separation. Microflow-LC is also being used for peptide bioanalysis to achieve higher sensitivity (Chen et al. 2019).

Targeting lower detection limits is an important challenge for peptide bioanalysis. Typically, LC-MS/MS quantitation of smaller peptides is done using MRM mode. However, with disulfide-bonded peptides, cyclic peptides, or other peptides that do not efficiently generate fragment ions under collision-induced, a pseudo-MRM mode is used to achieve sufficient sensitivity wherein the same ion is monitored in the Q1 and Q3 mode without fragmentation (Kang et al. 2020). Larger peptides tend to form multiple charge states during mass spectrometric ionization; the formation of multiple charge state species results in decreased signal response adversely impacting sensitivity. Charge state consolidation has been reported by addition of mobile phase additives such as dimethyl sulfoxide (DMSO) and m-nitrobenzyl alcohol to improve sensitivity (Kay et al. 2016; Nshanian et al. 2018). Also, hydrophobic peptides that do not ionize efficiently can be derivatized to improve mass spectrometric ionization.

6.3 Oligonucleotide Bioanalysis for Pharmacokinetic Evaluations

A typical ON therapeutic molecule ranges from 15 to 50 nucleotides long and can be an SS RNA or DNA, DS DNA, or a hybrid molecule. The ON molecule types are selected based on the desired mechanism of action. An ON length and sequence can be modulated to obtain target specificity and efficient mode of action (Roberts et al. 2020). In preclinical and clinical studies, the sample types and collection time points will be based on PK strategies. The sample types will include and not be limited to biofluids (plasma, CSF, and urine) and tissue types such as liver, kidney, spleen, bone marrow, and target organs, where possible. Methodologies chosen for ON bioanalysis will be based on the type and form of the ON. The bioanalysis involves evaluating the molecule itself and its metabolites in diverse sample types. The types of evaluations are based on the stage of drug development. Common platforms used to quantify ON molecules are LC-MS (liquid chromatography-mass spectroscopy) (Ewles et al. 2014) and LBA (ligand binding assay) (Wang and Ji 2016; Lorenson et al. 2019), and recently RT-qPCR-based method was developed for siRNA (Castellanos-Rizaldos et al. 2020). The RT-qPCR method is yet to be modified to quantify other types of ON molecules. These diverse assay platforms show very promising approaches to quantify ON with competitive sensitivity and specificity, but each have their pros and cons, which are illustrated in Table 6.1.

6.3.1 Assay Development Using LC-MS Platform

A typical LC-MS-based assay to quantitate ON comprises multiple steps. These steps can vary depending on the type of sample and measured ON molecule; a detailed list of steps is listed in the schematic diagram provided in Fig. 6.2. Irrespective of the sample or ON type, a sample processing step, before analyzing on an LC-MS instrument, is shown to be crucial. In every sample, a control ON called internal standard (IS) is added to show no effect on the therapeutic ON due to sample processing step. The IS is selected based on the criteria such as, not to interact with the therapeutic ON, different LC retention time, same length, backbone chemistry, and modification as the therapeutic ON molecule. The quantity of IS added to all the standards, QC samples, and test samples is based on the capacity of the processing steps (Zhang et al. 2007; Sips et al. 2019).

Assay development for SS ON needs steps ideal to extract full-length ON and its metabolites, without influencing the integrity of the oligo while processing it (Ewles et al. 2014). Similarly, DS siRNA molecules should be targeted to quantify full-length DS siRNA. So, sample processing steps should be carefully selected where DS siRNA and ASOs are unaltered. An ideal DS siRNA assay should be able to differentiate intact DS and its individual siRNA strands in the samples. The selection of sample processing steps will be based on characteristics of ON molecules

Table 6.1 Comparison of LC-MS and LBA platforms in ON quantification assays

Parameter	LC-MS	LBA	RT-qPCR
Sensitivity	Less sensitive with ranges >1 ng/mL in biofluids and >20 ng/g for tissues. The quantifiable range is about five orders of magnitude.	Better sensitivity at around 0.5 ng/mL for biofluids and <10 ng/g of tissue. The quantifiable range is about three to four orders of magnitude.	The drug concentration is interpolated using absolute quantification. Sensitivity at fg/mL to pg/mL. which is the most sensitive of the three available platforms.
Selectivity	Due to the extraction process selectivity is easy to pass.	Selectivity might have a challenge to pass due to a possible matrix interference.	Not enough information available.
Specificity	Extremely specific and can identify metabolites in addition to full length ON	Specific only to nucleotides truncated at 3' end of the ON	Needs thorough optimization to be specific.
Throughput	Low to medium throughput	High throughput is possible.	High throughput is possible.
Technical	Higher complexity while identifying ionized species specific to the full length.	Probes' design is crucial in performance of the assay. Ligation and nuclease enzyme performances could be dependent on the ON chemistries.	Enzyme activity for the PCR amplification step is crucial and can be affected by the ON chemistries.
Assay Cost efficiency	Assay Development costs are high. Advanced instruments such as HRMS, Triple-Quadrupole, etc. type LC/MS/MS are needed.	Relatively inexpensive assay costs. Regular spectrophotometer with a fluorescence detector is needed.	Relatively inexpensive assay costs. qPCR instruments are widely available in most of the molecular labs.

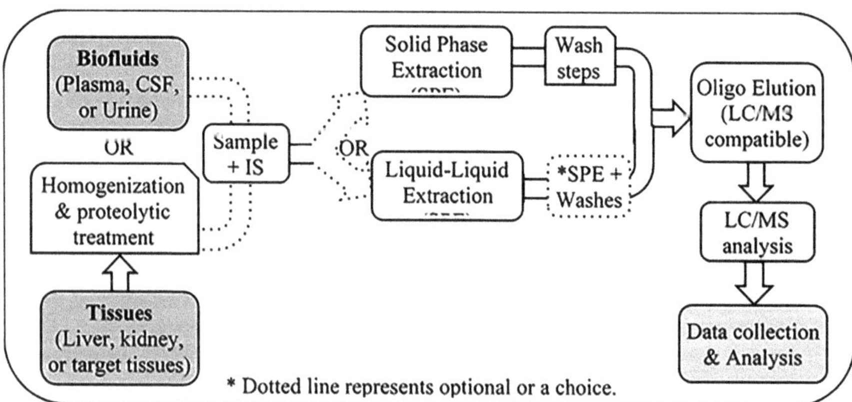

Fig. 6.2 Schematic diagram listing steps involving sample preparation and analysis on LC-MS platform

such as acidity, polarity, and protein binding nature of the molecules. In addition, it has been shown in vitro that >90% of the ON drug molecules are protein bound in plasma and tissue samples (Sips et al. 2019). During method development, especially in biofluid samples, ammonium precipitation, phenol-Chloroform extraction, or simple organic-phase extraction might be sufficient to separate proteins bound to the ON (van Dongen and Niessen 2011; Ewles et al. 2014). However, for tissue samples, proteinase K treatment will be included after optimization especially during the homogenization step for tissue treatment to remove the bound proteins (Nuckowski et al. 2018).

Widely used sample processing steps for ON-specific LC-MS methods are SPE (solid-phase extraction) (McGinley et al. 2010), liquid-liquid extraction (LLE) (Turnpenny et al. 2011; Nuckowski et al. 2018), and hybridization methods (Sips et al. 2019). These three methods will provide 80 to 90% efficiency in IS and ON recovery. Selection of these methods is completely based on the ON backbone chemistry, overall charge, expected concentration, adsorption, affinity, volume of sample to process, etc. SPE are fixed bed columns with various characteristics, which are available in single-vial or 96-well formats (McGinley et al. 2010; Nuckowski et al. 2018). On the other side, LLE methods use reagents such as 1-chlorobutane, ethyl acetate, hexane, and methyl tert-butyl ether (MTBE), or typical phenol-chloroform mixture with isopropyl alcohol can be used to extract ON (Ewles et al. 2014). Hybridization method is specifically used for SS ONs bioanalyses, where a nucleic acid probe that is complementary to the ON molecule is fixed in the substratum using biotin- streptavidin or other modifications (Sips et al. 2019). These extraction processes are used to process assay standards, quality control samples, and study samples.

Particularly, the SPE extraction process includes a few wash steps to clear any unwanted matrix interference, and the ON can be eluted in solvents compatible with LC-MS such as ammonia with acetonitrile. If a kit is used, a specific elution buffer compatible with the SPE material will be provided by the kit manufacturer (McGinley et al. 2010; Sips et al. 2019). The eluted sample from all the extraction processes can be injected and run on an LC-MS instrument. A choice of instruments is available to quantify the ON. The efficiency of ON quantification is dependent on three factors, which are chromatographic instrument's retention potential, efficient ionization and multiple negative charge state formations of ONs, and efficient fragmentation and exchange of phosphate groups. Typically used instruments are reverse-phase LC (RPLC) for SS ONs and soft ionization LC (IPLC) for DS DNA and siRNA molecules. The separation columns typically used are C4, C8, C18, and phenyl, which is mostly used on RPLC. Selection of these separation columns is based on the length and modifications of the ON. When a C4 or C8 column is used for ASO separation, the peaks are usually broadened, and the phenyl column gives a good separation of long ON molecules (Ewles et al. 2014; Hagelskamp et al. 2020).

Efficiency of ionization is crucial for an effective LC-MS method. Ionization efficiency of the ON will be dependent on the chromatographic conditions and MS parameters used while running it on the LC-MS instrument. The ion pairing reagents used for SS ON, siRNA, and dsRNA are TEAA, hexylammonium acetate, and TEA/

HFIP in combination with low-level MeCN gradients. Details of LC-MS conditions are provided in detail by van Dongen and Niessen (2011).

While developing the ON qualification assay, if sensitivity or selectivity is difficult to achieve to the level described in 2018 FDA bioanalytical guidance (FDA 2018a, b), different types of extraction methods listed above can be tested based on the ON characteristic for assay development and sample extraction. LC-MS has an advantage of specificity over the LBA platform. However, sensitivity of the LC-MS assay is less than other methods, such as LBA because of sample processing requirements and other differences between the platforms. Advanced LC-MS instruments such as SCIEX- Triple QUAD™ have promising technology to make the assays more sensitive and to make it a preferred platform for PK assays (Tozaki et al. 2018).

As metabolite stability, ON drugs are known to undergo enzyme processing. Endogenous enzymes such as endonucleases, exonucleases, ADAR, OAS1, RNASE L, etc. (Merkle et al. 2019; Roberts et al. 2020; Schwartz et al. 2020) can bind nascent ON molecules directly or when ON molecules are bound to the target region of RNA or DNA of the drug. It is important to quantify the full-length form of ON. However, for metabolite stability, the ON drug profile in the samples should also be identified. Sample extraction process for full-length ON and its metabolites can be identified from the same sample using the LC/MS procedures explained earlier.

6.3.2 Assay Development Using LBA Platform

Assays on ligand binding analytical platforms can be of two subtypes based on the detailed steps and the probe design for capture and detector probes to the ON molecule.

The first subtype is a simple nucleic acid sandwich-based hybridization ELISA (S-HELISA), where capture and detector probes are top half and the bottom half complementary sequences to the ON molecule. The probe lengths can be depending on the full-length ON. Capture probe is conjugated with free primary amine or biotin at 3′ end, and the 3′ end of detector probe is conjugated to digoxigenin, or sulfo-ruthenium or other tags used for detecting the sandwich formed. Detailed figure of the complex formed in L-HELISA is shown in Fig. 6.3b. If probes are not stable in matrix, certain nucleotides of the probes can be modified to locked nucleic acids (LNAs). These modifications can stabilize or inhibit hybridization due to any mismatches or change in melting temperature (T_m) (Levin et al. 2006; Mueller et al. 2018; Thayer et al. 2019). So, the probes with LNA modifications should be tested in both orientations as capture or detector probes. A carbon linker can be added to the conjugate to minimize steric hindrance. The higher the melting point of the probes, the tighter they bind with the ON molecules. The time it takes for assay development is shorter with satisfactory sensitivity and selectivity. However,

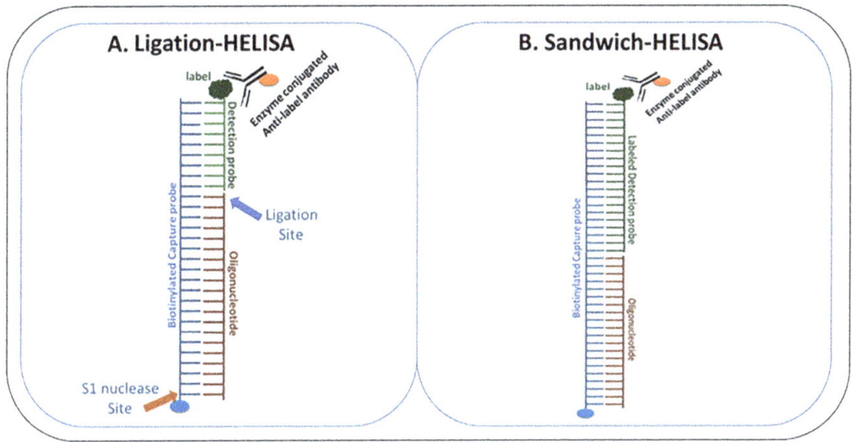

Fig. 6.3 Diagram showing complex of capture and detector probes, and ON to quantitate. (**a**) Ligation-HELISA and (**b**) Sandwich-HELISA. Arrow marks show the ligation and S1 nuclease sites, which are only feasible in ligation H-ELISA to provide certain specificity. Enzyme conjugate antibody specific to the label on the detector probe provides a quantifiable signal

specificity of the assay is very poor where even four nucleotide truncations on a side of the ON (n-4) can interfere with the assay. So, if metabolite profiling was done on an ON, presence of ≤ n-4 from both sides of the parent ON, the S-HELISA method is not feasible. In these situations alternate methods such as ligation HELISA or LC-MS method should be chosen over S-HELISA. Various tissue and biofluid samples can be tested using these methods with specificity of full-length ON. Samples are mixed with both the probes at MRD in a choice of assay dilution buffer, and pre-complexes of capture probe, ON, and detector probe are formed in sample matrix before added to a well capable to bind the complex using the capture probe, e.g., if the capture probe has biotin, appropriate streptavidin-coated plates should be used. The complexes are detected using an enzyme system such as anti-digoxigenin HRP or MSD reading buffer at manufacturer-recommended dilutions.

The second HELISA method is called ligation-mediated HELISA (L-HELISA) where the capture probe is complementary to the full-length ON and nine extra nucleotides that are different from the ON at the 5′ end. Detailed figure of the complex formed in L-HELISA is shown in Fig. 6.3a. The 3′ end of the capture probe is primary amine or biotin conjugated to bind to an appropriate 96-well plate such as amine activated or streptavidin MSD plates. The detector probe is conjugated to digoxigenin or sulfo-Ruthenium complementary to just the 9-mer at the 5′ end of the capture probe. The capture probe, ON, and detector can be formed on a 96-well plate in a sequential form or pre-made in a tube before adding on to an appropriate 96-well plate. To obtain specificity of the ON, the complexes will be treated with T4 DNA ligase and an S1 Mung bean nuclease in a stepwise fashion. The T4 DNA ligase provides specificity at the 3′ end of the ON facilitating the ligation of 3′ end of ON to 5′ phosphorylated end of the detector probe. If the 3′ of the ON is truncated by even a single nucleotide, the ligation of 3′ end of ON to 5′ phosphorylated end of the detector probe won't be successful. The un-ligated detector probe is washed away from the complex leading to no signal. Meanwhile, at the 3′ end of the capture, S1 Mung bean nucleotide facilitates in degrading the capture probe at 3′ end due to an overhang. However, due to steric hindrance, achieving 100% sensitivity at n-1 truncated at 5′ end of the ON is difficult. If a metabolic profile of an ON molecule has more than 10% of 5′ n-1 metabolite, the L-HELISA procedure is not suitable for ON quantification (Thayer et al. 2019). In these cases, if sensitivity is not a concern, LC-MS method should be developed and used to quantify full-length parent ON molecules specifically (Ewles et al. 2014).

6.3.3 Assay Development Using qPCR Platform

The therapeutic ON can also be quantified by molecular methods such as quantititative polymerase chain reaction (qPCR) methods. Currently, a few different methods were developed to harness sensitivity and quantification power of qPCR (Castellanos-Rizaldos et al. 2020; Shin et al. 2021). The performance of a qPCR method depends on the length of the ON. Recent work published by Castellanos-Rizaldos et al. (2020) shows that a threshold of a qPCR-based assay is with oligos that are ≥23. As mentioned in the beginning of the chapter, usual ON drugs are between 18 and 30 nucleotides. In most of the cases, length of an ON chosen as a therapeutic agent depends on its mode of action. Most of siRNA-based therapeutics are predominantly 25 nucleotides long (Castellanos-Rizaldos et al. 2020), and adenosine deaminase acting on RNA (ADAR)-based ASO therapeutics are predominantly 28 nucleotides long (Merkle et al. 2019). These two types of ON are RNA-based ON. So, quantifying these two types of ON might be feasible using Rt-qPCR assays. For these ONs, an additional step to reverse transcription should be performed to prepare complementary DNA (cDNA). The cDNA preparation is an important step before quantifying using it as a template for qPCR.

The methods published in an abstract authored by Shin et al. (2021) consist of two probes. The first probe has a 3′ region complement to half of the ON, and the second probe has a '5 region complementary to the remaining half of the ON. Remaining regions of the probe are primer and quencher binding regions. In the first step, the ON in the sample binds to the edges of the probe that are complementary to the ON and ligated using a ligase. This ligated nucleic acid double strand would be a template for downstream reactions to quantify the ON using qPCR. The siRNA concentrations ranged over 6 logs (1 fM to 1 μM), including the anticipated quantification range. The qPCR reaction had a calculated efficiency of 102%, based on the slope of −3.27.

6.3.4 Regulatory Requirements for Bioanalysis of Oligonucleotides

There are multiple platforms available for ON bioanalysis in preclinical or clinical study samples. Regulatory requirements should be strictly imposed when the program is in the development stage where bioanalysis should be carried out in IND or CTA enabling studies or clinical studies. Depending on the stage of the program, drug concentration in certain sample matrices such as tissues can be qualified and tested in GLP-like assays instead of validated assays. General small- or large-molecule bioanalytical regulatory guidance can also apply to ON bioanalysis. For example, if LBA or LC-MS assay platforms are used, the most recent FDA-approved bioanalytical method validation guidance should be used.

6.4 Immunogenicity Risk Assessment for Peptides and Oligonucleotide Therapeutics

6.4.1 Product-Specific Factors

Product-specific factors may increase or decrease the potential for, and the risk associated with, an immune response. Immunogenicity testing should be considered when changes are made to product-specific factors. Product-related factors influencing the immunogenicity of therapeutic peptides or ON include the origin and nature of the active substance (structural homology, endogenous modifications), modification of the native structures (such as modified amino acids or nucleotides), product and process related impurities (such as breakdown products, aggregates, etc.,), and formulation.

6.4.1.1 Amino Acid or Nucleotide Modifications

Any variations in the amino acids or nucleotides, and structural or analog changes to molecules results in physical, chemical, enzymatic degradation, and/or modification (e.g., deamidation, oxidation and sulfation). These changes during all the steps of manufacturing process and changes during storage may trigger an immune response.

6.4.1.2 Product Aggregation

Aggregation of peptides and ON might either reveal new epitopes or lead to the formation of multivalent epitopes, which may stimulate the immune system. Factors, which could be considered to contribute to aggregate formation, include formulation, purification processes, and storage conditions of intermediates and the finished product. There is evidence that higher-molecular-weight aggregates and particles are more potent in eliciting immune responses than lower-molecular-weight aggregates (Dintzis et al. 1989; Bachmann et al. 1993; Joubert et al. 2012) by either extensive cross-linking of B-cell receptors (Dintzis et al. 1989; Bachmann et al. 1993) or by enhancing antigen uptake, processing, and presentation (Seong and Matzinger 2004) peptides to T-cell for generation of high-affinity, isotype-switched IgG antibody, which is most often associated with neutralization of product efficacy. Thus, the use of methods for assessment of aggregates provides a robust measure of protein aggregation.

6.4.1.3 Impurities

Adjuvant activity can arise through multiple mechanisms, including the presence of microbial or host-cell-related impurities in therapeutic peptide products and any conjugates to the ON molecules (Verthelyi and Wang 2010; Eon-duval et al. 2012; Lau and Dunn 2018; Lee et al. 2019; Roberts et al. 2020). These innate immune response modulating impurities (IIRMIs), including lipopolysaccharide, β-glucan, and flagellin, high-mobility group protein B1 (HMGB1), and nucleic acids, exert immune-enhancing activity by binding to and signaling through toll-like receptors or other pattern-recognition receptors present on B-cells, dendritic cells, and other antigen-presenting cell populations (Verthelyi and Wang 2010). This signaling prompts maturation of antigen-presenting cells and/or serves to directly stimulate B-cell antibody production. Because even trace levels of IIRMIs can modify the immunogenicity of a therapeutic protein product, the assays used to detect them should have sensitivities to assess levels that may lead to clinically relevant immune responses.

6.4.1.4 Formulation

Formulation components are principally chosen for their ability to preserve the native conformation of the therapeutic molecules in storage by preventing denaturation due to hydrophobic interactions, as well as by preventing chemical degradation, including truncation, oxidation, and deamidation (Wakankar and Borchardt 2006). The stability of the formulation and the composition and the source of excipients may alter immunogenicity of therapeutic proteins and should be considered as possible causes of such events.

6.4.2 Immunogenicity Assessment for Peptide Therapeutics

During therapeutic peptide product development, elucidation of a specific underlying immunologic mechanism for immunologically related adverse events can facilitate the development of strategies to help mitigate their risk. Within the industry, the assessment of unwanted immunogenicity can be improved by using prediction tools such as in silico and in vitro methods, optimizing the performance of immunogenicity assays and learning from the clinical impact of other therapeutics that have already been administered to patients. In silico methods identify the number and location of T-cell epitopes able to bind Human Leukocyte Antigen (HLA) class II molecules with high affinity. In most cases, since the development of a mature IgG response implies underlying antigen-specific helper T-cell involvement, further in vitro and in vivo models are being used to confirm the potential of predicted T-cell epitopes to induce an immune response. Thus, using relevant in vitro cell-based assays, immunogenicity risk potential of therapeutic protein can be assessed by immune cell activation, proliferative responses, and cytokine release syndrome. Avoidance of T-cell epitopes present within therapeutic proteins is important in generating effective protein therapeutics with reduced immunogenicity (Chester et al. 2005).

PBMC (peripheral blood mononuclear cells)-based assays are currently the closest representation to the human immune system prior to the first human clinical trial. For the in vitro assays, there is a range of cellular assay formats (such as HLA-binding assays, DC phenotyping assays for assessing innate response, DC-T-cell proliferation assays, ELISPOT) to confirm the capacity of the predicted epitopes to elicit an immune response. Additionally, in vivo methods using several transgenic animal models may support and reveal some mechanisms underlying drug immunogenicity (Brinks et al. 2011).

6.4.2.1 In Silico Immunogenicity Analysis

An in-silico T-cell epitope screening platform is used for the identification of potential epitopes in protein and antibody targets (Jawa et al. 2020). It determines structural characteristics of the HLA receptor and binding affinities to predict potential peptide/HLA binding, a condition necessary for T-cell activation. During in silico analysis, a protein sequence is first parsed into overlapping 9 to 10-mer peptide frames, each of which is then evaluated for binding potential to each of common class II HLA alleles that "cover" the genetic backgrounds of most humans worldwide. Normalization of allele specific scores makes it possible to compare scores of any peptide frames across multiple HLA alleles and enables immunogenicity prediction on a global scale. By calculating the density of high-scoring frames within a protein, it is possible to estimate a protein's overall immunogenicity. Although in silico modelling may help identify T-cell epitopes, it does not predict whether immunogenicity will occur. Often in silico assays lead to an overestimation of the potential immunogenic T-cell epitopes, as not all peptides that fit into the HLA class II groove are generated by protein processing in vivo or stimulate T-cell responses. Furthermore, in silico approaches for prediction of peptide binding to MHC molecules cannot easily reveal which epitopes are the most naturally immunogenic and thus the most appropriate for inclusion or as biomarkers for immune monitoring.

6.4.2.2 MHC-Peptide Binding Assay

MHC-peptide binding assay determines the ability of each candidate peptide to bind to one or more MHC class II alleles. By comparing the binding to that of a high-affinity T-cell epitope, the most likely immunogenic peptides in a protein sequence can be identified. The protein sequence is broken down into an overlapping peptide library, and the ability of these peptides to bind to HLA molecules is assayed in vitro which can determine if the peptide could be immunogenic over a significant percentage of the population. Both in silico and MHC-peptide binding methods can be used to identify peptides that have strong affinity for HLA Class II haplotypes which enables them to identify candidate epitopes.

6.4.2.3 Transient Cell-Line-Based Reporter Assays

Therapeutic proteins can contain multiple impurities. Depending on the cell substrate and the manufacturing process, numerous other innate immune response modulating impurities (IIRMI) can be present. Such impurities, even when present at trace levels, have the potential to activate innate immune cells in peripheral blood or embedded in tissues causing expression of cytokines and chemokines, increasing antigen uptake, facilitating processing and presentation by antigen presenting cells, and fostering product immunogenicity. Cell lines transfected with Toll-like receptors (TLR) (RAW-Blue and THP-1) allow for the detection of a broader subset of IIRMIs and receptor-specific agonists. This method is sensitive to trace levels of IIRMI and provides information of the type of IIRMIs present in the product. Alternatively, the use of a combination of macrophage cell lines of human and mouse origin allows for the detection of a broader spectrum of impurities and could be used to screen products for the presence of IIRMIs and inform immunogenicity risk assessments, particularly in the context of comparability exercises between synthetic peptide and approved reference listed drugs.

6.4.2.4 Functional in Vitro Cell-Based Assay

Functional in vitro assays are most commonly used for T-cell epitope discovery. Fully characterized PBMC from healthy donors in terms of high-resolution HLA- and HLA-II allotypes are normally used for in vitro assays. PBMC isolated from each donor are cryopreserved under liquid nitrogen in multiple aliquots. Dendritic Cell-T (DC-T) cell assay is typically used to test samples that display biological activity which directly modulates CD4^{+} T cell activation, proliferation, and cytokine release. The DC-T assay helps identify the presence or absence of potential T-cell epitopes within proteins. For DC-T cell assays, non-depleted PBMC are used from which monocyte-derived dendritic cells (DC) and CD4 T cells are isolated. The recognition of these antigens requires the presence of an APC that is capable of processing and presenting peptides derived from the antigen (Gaubin et al. 1996).

Human monocyte-derived dendritic cells can be manipulated in vitro to model antigen processing by professional APC in vivo. T-cell proliferation in response to stimulation by a peptide-MHC complex can be measured by the dilution of a fluorescent dye, carboxyfluorescein succinimidyl ester (CFSE), that decreases in fluorescence intensity by half with each round of cell division and can be measured by flow cytometry. In addition to CFSE labeling, cells can be co-stained for expression of other cell surface markers, transcription factors, and/or intracellular cytokines that distinguish T-helper-cell phenotypes, including Th1, Th2, Th17, etc. (Pala et al. 2000; Basu et al. 2013). Proliferation assays are sensitive and low in cost and can be designed to assess phenotype of the responder cells. DC-T cell assay is typically used to test samples that display biological activity which directly modulates $CD4^+$ T cell activation, proliferation, and cytokine release.

Dendritic cell-T (DC-T) cell assay not only helps identify the presence or absence of potential T-cell epitopes within proteins but also determine "relative antigenicity" between structurally similar molecules that are comparable in their application, formulation, mode of action, and route of exposure. This assay assists identification of those proteins by assessing upregulation of DC phenotypic/activation markers such as MHC class II, CD86, and CD83 and to compare the magnitude of helper T-cell proliferative responses, therefore potentially resulting in the development of anti-drug antibodies (ADA). Furthermore, it can also be used for assessing the impact on antigenicity of non-protein factors such as post-translational modifications, degradation products, chemical entities given in combination therapies, and other parameters related to manufacturing processes, drug formulation, and stability.

The consequences of immune responses to therapeutic protein products can range from no apparent effect to serious adverse events, including life-threatening complications such as anaphylaxis, neutralization of the effectiveness of lifesaving or highly effective therapies, or neutralization of endogenous proteins with non-redundant functions. Thus, using relevant in silico and in vitro cell-based assays, immunogenicity risk potential of therapeutic protein can be assessed in generating effective protein therapeutics with reduced immunogenicity.

6.5 ADA Assay for Peptides and Oligonucleotide Therapeutics

ADA assay methodology used for peptide and ON is similar to the ADA testing in antibody drugs and gene therapy modalities, where in preclinical and clinical studies, serum or plasma is collected at various time points and analyzed for ADAs. Details of sample collection time points are explained in Chap. 5, ADA sections. ADA assay development requires quality critical reagents, such as positive controls and sensitive detection systems. Raising positive controls for both peptide and ON therapeutics can be challenging since the antibody titers in rabbits and other species are lower compared to large proteins. So, antigen for immunizing animals should be the drug molecules conjugated to a carrier protein, such as keyhole limpet haemocyanin (KLH). The purity of the conjugate can influence the drug-specific immune response.

Typical ADA assay for peptides can be on an MSD or colorimetric ELISA platform, where amine or biotinylated peptide can be used as a capture molecule by coating a 96-well plate. The detector peptide is conjugated to a fluorophore. In certain cases, sensitivity of this assay scheme can be challenged due to 1:1 stoichiometry of fluorophore to peptide. To overcome sensitivity issues, other formats that can amplify signal such as streptavidin to biotin detector system and antibody to enzyme system with higher stoichiometry, can be tried.

For ON ADA assays, a capture probe can be a drug molecule with amine at 5′ or 3′ end of the ON molecule. To capture all antibodies in a sample, a mixture of ON with 5′ or 3′ amine or biotin should be coated on a 96-well plate, capable of binding amines or biotin. Usually, the detector system should be HRP conjugated protein-AG. Another ON molecule as a detector may not provide sensitivity required. Protein-AG HRP can detect IgGs from various preclinical species and humans. So, this assay scheme can be used for detecting ADAs. Similar to peptide ADA assays, sensitivity can be challenging with ADA assays for ON. In certain cases, in addition to stoichiometry seen in peptide ADA assays, ON sensitivity might also be due to inability to isolate affinity purified positive control. In a situation where affinity purified positive control is not available, total IgG from immunized animals can be isolated and tested for assay sensitivity.

6.6 Regulatory Perspective

In response to concerns about the potential side effects of immune responses to the peptide drugs, regulatory bodies such as the FDA and the EMA have begun to request that each protein therapeutic be accompanied by an immunogenicity risk assessment (Wadhwa and Thorpe 2010; Wadhwa et al. 2015). For example, the recent EMA guidance mentions "predictive immunogenicity" as an approach sponsors could consider in their preclinical studies (EMA-CHMP 2007). Given the contribution of T-cell responses to the development of a detrimental anti-drug antibody (ADA) response and the emerging suite of tools for predicting T-dependent immunogenicity, FDA outlines and recommends adoption of a risk-based approach for evaluating and mitigating responses to therapeutic protein products that affect their safety and efficacy (FDA 2014; FDA 2019).

Similar FDA guidance on synthetic peptides – "ANDAs for Certain Highly Purified Synthetic Peptide Drug Products That Refer to Listed Drugs of rDNA Origin (2017)" – also recommended to demonstrate that the active ingredient in a synthetic peptide drug product is the "same" as the active ingredient in a previously approved peptide of rDNA origin. Differences in impurities, particularly peptide (product)-related impurities, may affect the safety or effectiveness of a peptide drug product. For justification for any new peptide-related impurity found at levels $\geq$0.10% and $\leq$ 0.5% of the drug substance that is not present in the reference listed drugs (RLD), assays need to demonstrate that each new impurity does not contain T-cell epitopes that have an increased affinity for major histocompatibility complex (MHC) molecules and that the proposed synthetic peptide does not alter the innate immune activity. Additionally, functional immune assays (in vivo or in vitro) need to be considered to support the absence of increased risk of immunogenicity potential of the drug product (formulated with synthetic drug substance) as compared to the reference listed drugs (RLD).

Regulatory requirements for ON are limited or not currently available. So, it will not be discussed in this chapter.

6.7 Future Perspective

Peptide and ON therapeutics are not new to the therapeutic world. However, due to multiple new drug approvals in these modalities, in the past decade, multiple biotechnological and pharmaceutical companies have established to bring new chemistry and structural changes to these molecules resulting in expanded druggable targets. These new molecules bring new challenges to the drug development, especially to bioanalytical methods. With the new molecules and new challenges, there is a chance for new bioanalytical methodologies and instruments developed assisting quantifying drug molecules, ultimately resulting in regulatory changes to provide ample guidance for drug development and new drug approvals.

Disclaimer Any opinions or forward-looking statements expressed are those of the authors and may not reflect views held by their employers (Thermo Fisher Scientific for Morse Faria, Spark therapeutics for Inderpal Singh, and Wave Life Sciences for Ramakrishna Boyanapalli).

References

Bachmann MF, Rohrer UH, Kundig TM, Burki K, Hengartner H, Zinkernagel RM. The influence of antigen organization on B cell responsiveness. Science. 1993;262(5138):1448–51.

Bajan S, Hutvagner G. RNA-based therapeutics: from antisense oligonucleotides to miRNAs. Cell. 2020;9:1.

Basu R, Hatton RD, Weaver CT. The Th17 family: flexibility follows function. Immunol Rev. 2013;252(1):89–103.

Bodera P, Stankiewicz W, Kocik J. Synthetic immunostimulatory oligonucleotides in experimental and clinical practice. Pharmacol Rep. 2012;64(5):1003–10.

Brinks V, Jiskoot W, Schellekens H. Immunogenicity of therapeutic proteins: the use of animal models. Pharm Res. 2011;28(10):2379–85.

Bronsema KJ, Bischoff R, van de Merbel NC. High-sensitivity LC-MS/MS quantification of peptides and proteins in complex biological samples: the impact of enzymatic digestion and internal standard selection on method performance. Anal Chem. 2013;85(20):9528–35.

Castellanos-Rizaldos E, Brown CR, Dennin S, Kim J, Gupta S, Najarian D, Gu Y, Aluri K, Enders J, Brown K, Xu Y. RT-qPCR methods to support pharmacokinetics and drug mechanism of action to advance development of RNAi therapeutics. Nucleic Acid Ther. 2020;30(3):133–42.

Chang D, Kolis SJ, Linderholm KH, Julian TF, Nachi R, Dzerk AM, Lin PP, Lee JW, Bansal SK. Bioanalytical method development and validation for a large peptide HIV fusion inhibitor (enfuvirtide, T-20) and its metabolite in human plasma using LC-MS/MS. J Pharm Biomed Anal. 2005;38(3):487–96.

Chappell DL, Lassman ME, McAvoy T, Lin M, Spellman DS, Laterza OF. Quantitation of human peptides and proteins via MS: review of analytically validated assays. Bioanalysis. 2014;6(13):1843–57.

Charignon D, Spath P, Martin L, Drouet C. Icatibant , the bradykinin B2 receptor antagonist with target to the interconnected kinin systems. Expert Opin Pharmacother. 2012;13(15):2233–47.

Chen Z, Alelyunas YW, Wrona MD, Kehler JR, Szapacs ME, Evans CA. Microflow UPLC and high-resolution MS as a sensitive and robust platform for quantitation of intact peptide hormones. Bioanalysis. 2019;11(13):1275–89.

Chester KA, Baker M, Mayer A. Overcoming the immunologic response to foreign enzymes in cancer therapy. Expert Rev Clin Immunol. 2005;1(4):549–59.

de la Torre BG, Albericio F. Peptide therapeutics 2.0. Molecules. 2020;25:10.

Dintzis RZ, Okajima M, Middleton MH, Greene G, Dintzis HM. The immunogenicity of soluble haptenated polymers is determined by molecular mass and hapten valence. J Immunol. 1989;143(4):1239–44.

EMA-CHMP. Guideline on immunogenicity assessment of biotechnology-derived therapeutic proteins; 2007.

Eon-duval A, Valax P, Solacroup T, Broly H, Gleixner R, Strat CL, Sutter J. Application of the quality by design approach to the drug substance manufacturing process of an fc fusion protein: towards a global multi-step design space. J Pharm Sci. 2012;101(10):3604–18.

Ewles M, Goodwin L, Schneider A, Rothhammer-Hampl T. Quantification of oligonucleotides by LC-MS/MS: the challenges of quantifying a phosphorothioate oligonucleotide and multiple metabolites. Bioanalysis. 2014;6(4):447–64.

FDA, U. Guidance for industry—immunogenicity assessment for therapeutic protein products; 2014.

FDA, U. Bioanalytical method validation guidance for industry. US Department of Health and Human Services Food and Drug Administration Center for Drug Evaluation and Research and Center for Veterinary Medicine; 2018a.

FDA, U. Definition of the term "biological product" final regulatory impact analysis; 2018b.

FDA, U. Immunogenicity testing of therapeutic protein products—developing and validating assays for anti-drug antibody detection; 2019.

FDA, U. IND submissions for individualized antisense oligonucleotide drug products: administrative and procedural recommendations. Draft Guidance. 2021a.

FDA, U. Nonclinical testing of individualized antisense oligonucleotide drug products for severely debilitating or life-threatening diseases. Draft Guidance.pdf.; 2021b.

FDA, U. ANDAs for certain highly purified synthetic peptide drug products that refer to listed drugs of rDNA origin guidance for industry; 2021c.

Featherston AL, Kwon Y, Pompeo MM, Engl OD, Leahy DK, Miller SJ. Catalytic asymmetric and stereodivergent oligonucleotide synthesis. Science. 2021;371(6530):702–7.

Gaubin M, Autiero M, Houlgatte R, Basmaciogullari S, Auffray C, Piatier-Tonneau D. Molecular basis of T lymphocyte CD4 antigen functions. Eur J Clin Chem Clin Biochem. 1996;34(9):723–8.

Hagelskamp F, Borland K, Ramos J, Hendrick AG, Fu D, Kellner S. Broadly applicable oligonucleotide mass spectrometry for the analysis of RNA writers and erasers in vitro. Nucleic Acids Res. 2020;48(7):e41.

Haynes JJ, Jones H, Gibson D, Clark GT. Bioanalytical determination of unstable endogenous small peptides: RFRP3 and its metabolites in rat blood. Bioanalysis. 2011;3(7):763–78.

Hornung V, Guenthner-Biller M, Bourquin C, Ablasser A, Schlee M, Uematsu S, Noronha A, Manoharan M, Akira S, de Fougerolles A, Endres S, Hartmann G. Sequence-specific potent induction of IFN-alpha by short interfering RNA in plasmacytoid dendritic cells through TLR7. Nat Med. 2005;11(3):263–70.

Iwamoto N, Butler DCD, Svrzikapa N, Mohapatra S, Zlatev I, Sah D, Meena SSM, Lu G, Apponi LH, Frank-Kamenetsky M, Zhang JJ, Vargeese C, Verdine GL. Control of phosphorothioate stereochemistry substantially increases the efficacy of antisense oligonucleotides. Nat Biotechnol. 2017;35(9):845–51.

Jawa V, Terry F, Gokemeijer J, Mitra-Kaushik S, Roberts BJ, Tourdot S, De Groot AS. T-cell dependent immunogenicity of protein therapeutics pre-clinical assessment and mitigation-updated consensus and review 2020. Front Immunol. 2020;11:1301.

Joubert MK, Hokom M, Eakin C, Zhou L, Deshpande M, Baker MP, Goletz TJ, Kerwin BA, Chirmule N, Narhi LO, Jawa V. Highly aggregated antibody therapeutics can enhance the in vitro innate and late-stage T-cell immune responses. J Biol Chem. 2012;287(30):25266–79.

Judge AD, Sood V, Shaw JR, Fang D, McClintock K, MacLachlan I. Sequence-dependent stimulation of the mammalian innate immune response by synthetic siRNA. Nat Biotechnol. 2005;23(4):457–62.

Kang L, Weng N, Jian W. LC-MS bioanalysis of intact proteins and peptides. Biomed Chromatogr. 2020;34(1):e4633.

Kay RG, Howard J, Stensson S. A current perspective of supercharging reagents and peptide bioanalysis. Bioanalysis. 2016;8(3):157–61.

Kay RG, Roberts A. Bioanalysis of biotherapeutic proteins and peptides: immunological or MS approach? Bioanalysis. 2012;4(8):857–60.

Lau JL, Dunn MK. Therapeutic peptides: historical perspectives, current development trends, and future directions. Bioorg Med Chem. 2018;26(10):2700–7.

Lee AC, Harris JL, Khanna KK, Hong JH. A comprehensive review on current advances in peptide drug development and design. Int J Mol Sci. 2019;20:10.

Levin JD, Fiala D, Samala MF, Kahn JD, Peterson RJ. Position-dependent effects of locked nucleic acid (LNA) on DNA sequencing and PCR primers. Nucleic Acids Res. 2006;34(20):e142.

Li W, Zhang J, Tse FL. Strategies in quantitative LC-MS/MS analysis of unstable small molecules in biological matrices. Biomed Chromatogr. 2011;25(1–2):258–77.

Liczner C, Duke K, Juneau G, Egli M, Wilds CJ. Beyond ribose and phosphate: selected nucleic acid modifications for structure-function investigations and therapeutic applications. Beilstein J Org Chem. 2021;17:908–31.

Liu Y, Dodart JC, Tran H, Berkovitch S, Braun M, Byrne M, Durbin AF, Hu XS, Iwamoto N, Jang HG, Kandasamy P, Liu F, Longo K, Ruschel J, Shelke J, Yang H, Yin Y, Donner A, Zhong Z, Vargeese C, Brown RH Jr. Variant-selective stereopure oligonucleotides protect against pathologies associated with C9orf72-repeat expansion in preclinical models. Nat Commun. 2021;12:847.

Lorenson MY, Chen KE, Walker AM. Enzyme-linked oligonucleotide hybridization assay for direct oligo measurement in blood. Biol Methods Protoc. 2019;4(1):bpy014.

Maes K, Smolders I, Michotte Y, Van Eeckhaut A. Strategies to reduce aspecific adsorption of peptides and proteins in liquid chromatography-mass spectrometry based bioanalyses: an overview. J Chromatogr A. 2014;1358:1–13.

McGinley M, Scott G, Rivera B. Analyzing oligos from biological matrices. GEN Genetic Eng Biotech News. 2010;30:11.

Merkle T, Merz S, Reautschnig P, Blaha A, Li Q, Vogel P, Wettengel J, Li JB, Stafforst T. Precise RNA editing by recruiting endogenous ADARs with antisense oligonucleotides. Nat Biotechnol. 2019;37(2):133–8.

Montazersaheb S, Hejazi MS, Nozad Charoudeh H. Potential of peptide nucleic acids in future therapeutic applications. Adv Pharm Bull. 2018;8(4):551–63.

Mueller KT, Waldron E, Grupp SA, Levine JE, Laetsch TW, Pulsipher MA, Boyer MW, August KJ, Hamilton J, Awasthi R, Stein AM, Sickert D, Chakraborty A, Levine BL, June CH, Tomassian L, Shah SS, Leung M, Taran T, Wood PA, Maude SL. Clinical pharmacology of Tisagenlecleucel in B-cell acute lymphoblastic Leukemia. Clin Cancer Res. 2018;24(24):6175–84.

Muttenthaler M, King GF, Adams DJ, Alewood PF. Trends in peptide drug discovery. Nat Rev Drug Discov. 2021;20(4):309–25.

Nshanian M, Lakshmanan R, Chen H, Ogorzalek Loo RR, Loo JA. Enhancing sensitivity of liquid chromatography-mass spectrometry of peptides and proteins using supercharging agents. Int J Mass Spectrom. 2018;427:157–64.

Nuckowski L, Kaczmarkiewicz A, Studzinska S. Review on sample preparation methods for oligonucleotides analysis by liquid chromatography. J Chromatogr B Analyt Technol Biomed Life Sci. 2018;1090:90–100.

Pala P, Hussell T, Openshaw PJ. Flow cytometric measurement of intracellular cytokines. J Immunol Methods. 2000;243(1–2):107–24.

Polson C, Sarkar P, Incledon B, Raguvaran V, Grant R. Optimization of protein precipitation based upon effectiveness of protein removal and ionization effect in liquid chromatography-tandem mass spectrometry. J Chromatogr B Analyt Technol Biomed Life Sci. 2003;785(2):263–75.

Quijano E, Bahal R, Ricciardi A, Saltzman WM, Glazer PM. Therapeutic peptide nucleic acids: principles, limitations, and opportunities. Yale J Biol Med. 2017;90(4):583–98.

Rauh M. LC-MS/MS for protein and peptide quantification in clinical chemistry. J Chromatogr B Analyt Technol Biomed Life Sci. 2012;883-884:59–67.

Roberts TC, Langer R, Wood MJA. Advances in oligonucleotide drug delivery. Nat Rev Drug Discov. 2020;19(10):673–94.

Schwartz SL, Park EN, Vachon VK, Danzy S, Lowen AC, Conn GL. Human OAS1 activation is highly dependent on both RNA sequence and context of activating RNA motifs. Nucleic Acids Res. 2020;48(13):7520–31.

Seong SY, Matzinger P. Hydrophobicity: an ancient damage-associated molecular pattern that initiates innate immune responses. Nat Rev Immunol. 2004;4(6):469–78.

Shin M, Krishnamurthy PM, Watts JK. Quantification of antisense oligonucleotides by splint ligation and quantitative polymerase chain reaction. bioRxiv. 2021;2021:2006.2005.447195.

Shokrzadeh N, Winkler AM, Dirin M, Winkler J. Oligonucleotides conjugated with short chemically defined polyethylene glycol chains are efficient antisense agents. Bioorg Med Chem Lett. 2014;24(24):5758–61.

Sips L, Ediage EN, Ingelse B, Verhaeghe T, Dillen L. LC-MS quantification of oligonucleotides in biological matrices with SPE or hybridization extraction. Bioanalysis. 2019;11(21):1941–54.

Sutton JM, Kim J, El Zahar NM, Bartlett MG. Bioanalysis and biotransformation of oligonucleotide therapeutics by liquid chromatography-mass spectrometry. Mass Spectrom Rev. 2021;40(4):334–58.

Thayer MB, Lade JM, Doherty D, Xie F, Basiri B, Barnaby OS, Bala NS, Rock BM. Application of locked nucleic acid oligonucleotides for siRNA preclinical bioanalytics. Sci Rep. 2019;9(1):3566.

Thiel KW, Giangrande PH. Therapeutic applications of DNA and RNA aptamers. Oligonucleotides. 2009;19(3):209–22.

Thomas A, Schanzer W, Delahaut P, Thevis M. Immunoaffinity purification of peptide hormones prior to liquid chromatography-mass spectrometry in doping controls. Methods. 2012;56(2):230–5.

Thomas A, Schanzer W, Thevis M. Immunoaffinity techniques coupled to mass spectrometry for the analysis of human peptide hormones: advances and applications. Expert Rev Proteomics. 2017;14(9):799–807.

Tozaki T, Karasawa K, Minamijima Y, Ishii H, Kikuchi M, Kakoi H, Hirota KI, Kusano K, Nagata SI. Detection of phosphorothioated (PS) oligonucleotides in horse plasma using a product ion (m/z 94.9362) derived from the PS moiety for doping control. BMC Res Notes. 2018;11(1):770.

Turnpenny P, Rawal J, Schardt T, Lamoratta S, Mueller H, Weber M, Brady K. Quantitation of locked nucleic acid antisense oligonucleotides in mouse tissue using a liquid-liquid extraction LC-MS/MS analytical approach. Bioanalysis. 2011;3(17):1911–21.

van den Broek I, Sparidans RW, Schellens JH, Beijnen JH. Quantitative bioanalysis of peptides by liquid chromatography coupled to (tandem) mass spectrometry. J Chromatogr B Analyt Technol Biomed Life Sci. 2008;872(1–2):1–22.

van Dongen WD, Niessen WM. Bioanalytical LC-MS of therapeutic oligonucleotides. Bioanalysis. 2011;3(5):541–64.

van Midwoud PM, Rieux L, Bischoff R, Verpoorte E, Niederlander HA. Improvement of recovery and repeatability in liquid chromatography-mass spectrometry analysis of peptides. J Proteome Res. 2007;6(2):781–91.

Verthelyi D, Wang V. Trace levels of innate immune response modulating impurities (IIRMIs) synergize to break tolerance to therapeutic proteins. PLoS One. 2010;5(12):e15252.

Wadhwa M, Knezevic I, Kang HN, Thorpe R. Immunogenicity assessment of biotherapeutic products: an overview of assays and their utility. Biologicals. 2015;43(5):298–306.

Wadhwa M, Thorpe R. Unwanted immunogenicity: lessons learned and future challenges. Bioanalysis. 2010;2(6):1073–84.

Wakankar AA, Borchardt RT. Formulation considerations for proteins susceptible to asparagine deamidation and aspartate isomerization. J Pharm Sci. 2006;95(11):2321–36.

Wang J, Lon HK, Lee SL, Burckart GJ, Pisetsky DS. Oligonucleotide-based drug development: considerations for clinical pharmacology and immunogenicity. Ther Innov Regul Sci. 2015;49(6):861–8.

Wang L, Ji C. Advances in quantitative bioanalysis of oligonucleotide biomarkers and therapeutics. Bioanalysis. 2016;8(2):143–55.

Wu JC, Meng QC, Ren HM, Wang HT, Wu J, Wang Q. Recent advances in peptide nucleic acid for cancer bionanotechnology. Acta Pharmacol Sin. 2017;38(6):798–805.

Xu Y, Garofolo F, Musuku A. The exciting world of oligonucleotides: a multidisciplinary complex challenge for multitasking ingenious bioanalysts. Bioanalysis. 2019;11(21):1905–8.

Xu Y, Goykhman D, Wang M, Jackson T, Xie I, Willson K, Breidinger S, Schuck H, Harrelson J, Woolf EJ. Strategy for peptide quantification using LC-MS in regulated bioanalysis: case study with a glucose-responsive insulin. Bioanalysis. 2018;10(15):1207–20.

Xu Y, Mehl JT, Bakhtiar R, Woolf EJ. Immunoaffinity purification using anti-PEG antibody followed by two-dimensional liquid chromatography/tandem mass spectrometry for the quantification of a PEGylated therapeutic peptide in human plasma. Anal Chem. 2010;82(16):6877–86.

Xu Y, Sun L, Anderson M, Belanger P, Trinh V, Lavallee P, Kantesaria B, Marcoux MJ, Breidinger S, Bateman KP, Goykhman D, Woolf EJ. Insulin glargine and its two active metabolites: a sensitive (16pM) and robust simultaneous hybrid assay coupling immunoaffinity purification with LC-MS/MS to support biosimilar clinical studies. J Chromatogr B Analyt Technol Biomed Life Sci. 2017;1063:50–9.

Yuan L. Sample preparation for LC-MS bioanalysis of peptides. In: Sample preparation in LC-MS bioanalysis; 2019. p. 284–303.

Zhang G, Lin J, Srinivasan K, Kavetskaia O, Duncan JN. Strategies for bioanalysis of an oligonucleotide class macromolecule from rat plasma using liquid chromatography-tandem mass spectrometry. Anal Chem. 2007;79(9):3416–24.

Index

S. Kumar (ed.), *An Introduction to Bioanalysis of Biopharmaceuticals*, AAPS Advances in the Pharmaceutical Sciences Series 57,
https://doi.org/10.1007/978-3-030-97193-9

L

M

N

O

P

Q

R

S

T

GPSR Compliance
The European Union's (EU) General Product Safety Regulation (GPSR) is a set of rules that requires consumer products to be safe and our obligations to ensure this.

If you have any concerns about our products, you can contact us on

ProductSafety@springernature.com

In case Publisher is established outside the EU, the EU authorized representative is:

Springer Nature Customer Service Center GmbH
Europaplatz 3
69115 Heidelberg, Germany

www.ingramcontent.com/pod-product-compliance
Ingram Content Group UK Ltd.
Pitfield, Milton Keynes, MK11 3LW, UK
UKHW021834270726
14058UKWH00001B/146

9783030999988